Project AIR FORCE | **RAND**

AEROSPACE OPERATIONS
in URBAN ENVIRONMENTS

Exploring New Concepts

Alan Vick, John Stillion, David R. Frelinger, Joel Kvitky,
Benjamin S. Lambeth, Jefferson P. Marquis, Matthew C. Waxman

Prepared for the United States Air Force

Motivated by a recent spate of urban operations in Panama, Somalia, Haiti, and Bosnia, the Department of Defense is now putting considerable effort into identifying and correcting shortcomings in the United States' ability to successfully conduct urban military operations. The Joint Urban Working Group (UWG), sponsored by J-8, is writing new joint urban doctrine as well as identifying problems and potential solutions relating to individual equipment, communications, navigation, surveillance, and weapons employment. The U.S. Air Force (USAF) has taken a keen interest in the UWG and took the lead in preparing the *Handbook for Joint Urban Operations*, a document intended to provide Joint Force Commanders and their staffs a primer on urban operations, bridging the gap until joint doctrine for urban operations is published.

At the request of General Ralph Eberhart, then–USAF Vice Chief of Staff, and with the sponsorship of Major General Norton Schwartz, Director of Strategic Planning, Headquarters USAF, Project AIR FORCE undertook a year-long investigation of the role that aerospace forces can play in joint urban military operations. This study sought to help the USAF better understand how the urban physical, social, and political environment constrains aerospace operations, to identify key operational tasks that aerospace forces can help accomplish, and to develop new concepts of operation to enhance the contribution that aerospace forces make to joint urban operations. It builds on previous work that Project AIR FORCE has conducted on the role of aerospace forces in lesser conflicts:

- MR-697-AF, *Enhancing Air Power's Contribution Against Light Infantry Targets*

- MR-842-AF, *Preparing the U.S. Air Force for Military Operations Other Than War.*

This report presents the findings of our study. It should be of interest to Air Force personnel in operations, plans, intelligence, and acquisition organizations. It should also be of interest to aviators in the sister services. Finally, it is our hope that the report will help soldiers, marines, and sailors better appreciate the contribution that aerospace forces can make to joint urban operations. The report is intended as both a reference source and an informative piece of writing. For this reason, clarifying information may be repeated in different chapters.

This study was conducted as part of the Strategy and Doctrine Program in RAND's Project AIR FORCE.

PROJECT AIR FORCE

Project AIR FORCE, a division of RAND, is the Air Force's federally funded research and development center (FFRDC) for studies and analysis. It provides the USAF with independent analysis of policy alternatives affecting the deployment, employment, combat readiness, and support of current and future air and space forces. Research is performed in four programs: Aerospace Force Development; Manpower, Personnel and Training; Resource Management; and Strategy and Doctrine.

CONTENTS

Preface   iii

Figures   ix

Tables   xi

Summary   xiii

Acknowledgments   xxi

Abbreviations and Acronyms   xxv

Chapter One
   INTRODUCTION   1
   Background   1
   How Likely Are Urban Operations?   3
   The Nature of Urban Military Operations   6
      From Rubble to Routine   7
      Redefining Warfare, Not Service Roles   8
   Purpose   10
   Organization   11

Chapter Two
   USING AEROSPACE POWER TO PREVENT THE
      URBAN FIGHT   13
   Introduction   13
   How Aerospace Power Can Forestall Urban Operations . .   17
      Anatomy of an Urban Military Operation   18
      Comparative Advantages of Aerospace Power   19
      Preclusion Strategies   21

Where Preclusionary Strategies May Have Worked   25
  Al Khafji.   25
  Kuwait City   28
  Southern Lebanon   28
Where Aerospace Power Failed to Prevent Urban
  Combat   29
  Lessons Learned   32
  The Second Chechen Battle   33
When Circumstances Leave No Choice   34
  As a Tool for Joint Force Commanders   36
  The Limits of Aerospace Power in Urban Operations . . .   37
Summary   37

Chapter Three
LEGAL AND POLITICAL CONSTRAINTS ON URBAN
  AEROSPACE OPERATIONS   39
Introduction   39
Legal Constraints on Urban Aerospace Operations   40
  Reciprocal Legal Duties and the Defender's
    Obligations   46
  The International Legal Challenges of Urban
    Environments   47
Political Constraints on Urban Operations   52
  Sensitivity to American Casualties   53
  Sensitivity to Collateral Damage and Civilian
    Suffering   55
Restrictive Rules of Engagement and Targeting   57
The Asymmetry of Constraints   62
Conclusion   68

Chapter Four
AEROSPACE OPERATIONS AND THE URBAN PHYSICAL
  ENVIRONMENT   71
Introduction   71
The Nature of Urban Terrain   72
  Urban Terrain Zones and Their Characteristics   74
  Air Defenses on Urban Terrain   80
Surveillance and Reconnaissance in the Urban
  Environment   83
  Sensor Cueing in the Urban Environment and a New
    Cueing Concept   84

Target Identification/Tracking in the Urban
    Environment   89
EO Sensors Versus Targets on Rooftops and in
    Streets   93
EO Sensors Versus Targets Inside Buildings   98
Aerial Weapon Delivery in the Urban Environment   107
Aerial Weapon Trajectories and the Geometry of
    Urban Terrain   107
Weapon Effects and Urban Terrain   111
Airlift in Urban Environments   114
Conclusions   115

Chapter Five
NEW CONCEPTS FOR ACCOMPLISHING KEY TASKS
    IN URBAN OPERATIONS   119
Introduction   119
Stop Movement of Combatants, Vehicles, and
    Equipment   120
Provide Rapid, High-Resolution Imagery for Target ID . . .   123
Detect and Neutralize Adversary Ambush Positions   127
Detect and Neutralize Snipers   131
Monitor High-Priority Targets   138
Resupply Isolated Friendly Ground Forces   143
Provide Close Support for Ground Forces   145
The Role of the Joint Control Center   146

Chapter Six
ENABLING TECHNOLOGIES FOR URBAN
    AEROSPACE OPERATIONS   149
Introduction   149
Three-Dimensional Modeling of the Urban
    Environment   150
Key Functions and the Scope of Air Force
    Involvement   150
Laser Radar Mapping   152
Stereoscopic Electro-Optical Imaging   155
Interferometric Synthetic Aperture Radar   156
Comparison of 3-D Imaging Technology   161
Platform Trade-Offs   162
Software Exploitation of 3-D Urban Maps   163

Communications and Navigation Technology for the
  Urban Environment 165
  UAV Relays 165
  Through-the-Wall Communications 166
  Pseudolites 169
Imaging Sensor Technology for Urban Operations 170
Non-Imaging Sensor Technology for Urban Operations . . 174
  Seismic and Acoustic Sensors 174
  Through-the-Wall Radar 175
  Remote Listening 179
  Chemical Sniffing 180
Sensor Fusion in Support of Urban Operations 182
Air-Launched Sensor Platforms 184
Limited-Effects Munitions 187
  Kinetic-Energy Weapons 187
  "Laser-Guided Hand Grenades" 188
  Miniature Glide Bombs, Cruise Missiles, and
    Killer UAVs 188
  Nonlethal Weapons 189
Conclusion 197

Chapter Seven
  CONCLUSIONS 199
  Introduction 199
  Key Findings 199
  The Need for Improved Urban Operational Capabilities . . 201
  Technological Promises and the Reality of War 202
  Next Steps 204

Appendix
A. TRIGONOMETRIC CALCULATIONS 207
B. MICROWAVE RECHARGING OF MINI-UAVs AND
    MICRO-UAVs 209
C. DETECTING SNIPERS 213
D. LESSONS LEARNED FROM PAST URBAN AIR
    OPERATIONS 217

Bibliography 267

FIGURES

1.1. Analytic Dimensions of Urban Military Operations . . .   9
3.1. Force Protection Versus Collateral-Damage
     Avoidance   59
4.1. Building Height and Street Width Affect
     Required Observer Height   74
4.2. Proportion of Surveyed Cities in Each UTZ   77
4.3. Surveillance/Reconnaissance Platform Coverage
     on Flat, Open Terrain   84
4.4. Surveillance/Reconnaissance Platform
     on Urban Terrain   86
4.5. Urban Sensor-Cueing Concept   88
4.6. Airborne Sensor View of Three-Fourths of 15-m-Wide
     Streets over Buildings of Different Heights, from
     1,000 m AGL   95
4.7. Targets in Buildings Allow Lower Depression
     Angles Than Do Targets in Streets   99
4.8. Perceived Window Size As a Function of Vertical
     Viewing Angle 101
4.9. Interior Views Shift with Vertical Viewing Angle 103
4.10. Effect of Azimuth and Elevation on Perceived
      Window Size 104
4.11. View of Building Faces Across 15-m-Wide Streets 105
4.12. Standard LGB Deliveries for Vertical and
      Horizontal Targets 109
4.13. Percent of City Wall Areas That Can Be Hit 110
4.14. Large USAF Warheads Make Damage Limitation
      Difficult for Many Urban Targets 113

5.1. Sensors Detect and Stop Combatant Force
      Movement or Operations   123
5.2. GPS-Guided Video Camera Is Used for Target ID   126
5.3. Pattern Analysis Detects Anomalies, Warns of
      Ambushes.   131
5.4. Armed UAV and Unmanned Ground Sensors
      Counter Snipers   135
5.5. Mini–Glide Bomb Flyout   137
5.6. Schematic of Mini–Glide Bomb   138
5.7. Covert Placement of Rooftop Unattended Sensor   141
5.8. Unattended Ground Sensors, SOF Team, and UAVs
      Monitor High-Priority Target (Covert)   142
5.9. GPS-Guided Canister Resupplies Friendly Forces   144
5.10. Attack Mini-UAV Provides Close Support   145
6.1. High-Resolution Laser Radar–Derived Digital
      Elevation Model of San Francisco   154
6.2. Single-Pass InSAR Geometry   157
6.3. InSAR Combined with Panchromatic Image Overlay
      of Howard Air Force Base, Panama   158
6.4. Layover and Shadowing Effects in SAR Imagery   159
6.5. Imaging-Sensor Weight and Performance
      Trade-Offs   172
A.1. Determining Maximum Horizontal Distance
      for Viewing Streets over a Building   207
B.1. Irradiance from a 24-ft^2 Antenna at 60,000 ft   211
C.1. Free-World Countersniper Systems   214

TABLES

S.1. Operational Tasks and Concepts of Operation   xviii
4.1. Urban Terrain Zone Classification System   76
4.2. Building Attributes by Urban Terrain Zone   79
4.3. UAV Dimensions   91
4.4. Average UTZ Areas, Attributes, and UAVs Required
     for Line of Sight to Any Part of Each UTZ   96
4.5. Building Attributes by Urban Terrain Zone 100
4.6. Current and New-Concept Warhead Weights
     and Effects 112
B.1. Sample Characteristics of Mini-UAVs and MAVs 210

SUMMARY

Since the end of the Cold War, U.S. forces have been involved in a number of operations (peacekeeping, humanitarian relief, and non-combatant evacuations) that have taken place in urban settings. Peace operations in Somalia, especially the deaths of 18 U.S. servicemen and the wounding of almost 100 others on October 3, 1993, profoundly influenced the American public's perceptions of modern urban combat in the developing world. For military professionals, Somalia also was a painful reminder that the technological and operational dominance the United States experienced on the conventional battlefield during Desert Storm did not necessarily carry over into urban peacekeeping. For infantrymen in particular, the fierce fighting of "Bloody Sunday"—the most intense light infantry engagements since the Vietnam War—brought home the relevance of urban combat, its nastiness, and the need to develop concepts and tactics better suited to this unique environment.

This report assesses the likelihood that the U.S. military will be called upon to conduct urban operations; explores the political, legal, and physical aspects of the urban operational environment; presents new concepts to accomplish key operational tasks; identifies key technologies that will need to be developed to execute these concepts; and offers lessons from past operations.

KEY FINDINGS

Key findings of this study are as follows:

- Global urbanization, particularly in the developing world, makes it highly likely that many—if not most—future military operations will have an urban component (although not necessarily one involving fighting).

- However, an increase in urban operations does not mean that conflict has become primarily an urban phenomenon or that non-urban military operations have been eclipsed. Rather, built-up areas are yet another environment in which military forces must be prepared to operate.

- Urban areas, with their physical and social complexity, are extremely difficult to operate in. Where possible, U.S. forces should avoid them. Aerospace forces can help preclude some urban military operations through deterrence, early warning, and rapid humanitarian or military intervention. Along with ground-based long-range fires, aerospace forces can also interdict adversary forces, potentially preventing them from reaching urban areas.

- Where urban operations cannot be avoided, aerospace forces can make important contributions to the joint team (air, land, sea, and space forces working together): detecting adversary forces in the open; attacking them in a variety of settings; and providing close support, navigation and communications infrastructure, and resupply for friendly ground forces.

- Offboard sensors for manned aircraft, three-dimensional urban mapping, Global Positioning System (GPS) relays on unmanned aerial vehicles (UAVs), and limited-effects weapons[1] have the potential to enhance the ability of aerospace forces to detect and attack adversary forces when rules of engagement are highly restrictive, such as in peace operations, noncombatant evac-

---

[1]*Limited-effects weapons,* or *munitions,* are designed to incapacitate or kill personnel targets without harming nearby civilians or friendly forces. Small, slow-moving weapons with grenade-like explosives or nonlethal warheads will be necessary to achieve this goal.

uations, and humanitarian assistance. Their development should be encouraged.

- Three-dimensional mapping and GPS relays also have the potential to substantially improve the situational awareness of friendly ground forces, allowing the smallest units as well as their commanders to know their location (both GPS coordinates and position in buildings). Coupling these technologies with laser rangefinders should allow friendly forces to quickly map the location of engaged adversary forces.

- Automated integration and pattern analysis of inputs from large, multiphenomenology sensor networks (i.e., sensors that use acoustic, infrared, seismic, chemical, and radar detectors) will be necessary to interpret the massive volume of activity found in most urban areas.

- But, in the type of limited operations this report emphasizes, we think it *unlikely* that automated classification of weapons, adversary personnel, or vehicles will be sufficiently reliable to permit lethal fires to be put automatically on targets. Rather, we expect that practical limitations of automated fusion, coupled with political concerns about collateral damage and civilian casualties, will dictate at least one human decisionmaker remaining in the loop between sensor and shooter.

- As long as human decisionmakers remain in the loop between sensor and shooter, human-machine interfaces will be a critical information-architecture issue. A major challenge will be developing the organizational processes that make quick decisions possible in light of the likely uncertainty and ambiguity associated with real combat. Without a responsive and agile command and control system, an elusive and adaptable adversary is likely to be there and gone before weapons can be brought to bear.

THE NEED FOR IMPROVED URBAN OPERATIONAL CAPABILITIES

Are urban-centered conflicts becoming more common? Are they a new form of warfare that will supplant traditional maneuver warfare in the open? These intriguing questions deserve serious and careful consideration by defense planners and researchers alike. At this point, there is insufficient evidence or analysis to answer these questions. Defense planners—who must walk a narrow path between apocalyptic and complacent visions of the future security environment—should focus, at least for now, on ensuring that the U.S. military can meet a broad range of urban-operation challenges, whether in major wars or small-scale operations.

The best argument for improved capabilities for urban operations is that, despite its best efforts to avoid cities, the U.S. military has had to fight in them in diverse circumstances. As well, in an increasingly urbanized world, noncombatant evacuations, humanitarian relief, and other noncombat operations are likely to take place in urban settings. The military is tasked in many such operations, because armed interference is a possibility. Although U.S. forces have usually been able to avoid combat during these operations, they must be prepared to conduct urban evacuations and humanitarian relief in the face of armed opposition. In short, whether in conventional conflicts or in smaller-scale contingencies, there is a good chance that U.S. forces will be called upon to operate in urban settings. The probability or desirability of urban operations need not be overstated to acknowledge that prudent defense planning requires that the United States develop the doctrine, training, organizations, equipment, and concepts of operations to be effective in this unique and difficult environment.

Planners also need to distinguish clearly between the problem of conducting military operations in the midst of a civilian population and that of fighting in the rubble of a largely abandoned city. The former problem is more complex and deserves analytic attention. As the Department of Defense (DoD) places more emphasis on peace, relief, counterterrorism, and other operations at the lower end of the conflict spectrum, planners, operators, and analysts must all gain a deeper understanding of the human and physical intricacies of the urban environment.

NEW CONCEPTS

Owing to their political, legal, and military complexity, operations at the lower end of the spectrum will be the most common and challenging urban missions the USAF will face. Although the USAF can improve its ability to conduct urban operations against conventional foes, the major shortcomings will be in situations where unconventional foes are conducting limited operations; therefore, we present new concepts directed at those situations.

It may be possible in future limited operations to identify and attack critical adversary centers of gravity or key nodes. However, historical experience suggests that this is the exception rather than the rule. In most limited conflicts, strategic objectives are more likely to be achieved through the cumulative effect of persistent surveillance and strike than through the destruction of a fairly small target set (e.g., an adversary's headquarters).

For example, in a notional peace operation that included an urban component, U.S. objectives might be to stop the violence, resettle the population, and achieve a return to normal in which routine civil and economic activities could take place without disruption. To accomplish these objectives would require, above all else, that friendly forces control the streets. The operational task of controlling the streets would, in turn, require that a variety of tactical tasks be accomplished. Table S.1 lists some of the more prominent tasks, along with the concepts of operation we propose for their accomplishment.

NEXT STEPS

The fielding of new urban surveillance, strike, and navigation technologies, when combined with innovative concepts of operation, could yield a significant improvement in USAF and joint capabilities for urban operations. In particular, the integration of ground sensor networks, low-flying air-launched UAVs, and more-traditional surveillance platforms with other platforms carrying limited-effects weapons could make joint forces much more effective in constrained urban operations. Developing the ability to detect, identify, and eliminate room-sized targets without collateral damage is a natural step in the ongoing evolution of aerospace power, simply continuing current trends in command, control, communication, intelligence, surveillance, and reconnaissance (C3ISR), battle management, and precision strike.

Table S.1

Operational Tasks and Concepts of Operation

Operational Task	Concept of Operation
Stop movement of combatants, vehicles, equipment	Air-dropped lifting-body sensor with parafoil extends eyes of manned platforms
Provide rapid high-resolution imagery for target ID	Offboard sensors flying at lower altitudes aid combat aircraft in identifying targets
Detect and neutralize adversary's ambush positions	Pattern analysis for anomaly detection
	RF resonance detector on UAV
Detect and neutralize snipers	Passive infrared sensor on UAV detects bullet in flight; glide bomb kills sniper
Monitor high-priority targets	UAVs–ground sensor–special operations force (SOF) team provide high-resolution imagery
Resupply isolated friendly ground forces	A GPS-guided canister of prepackaged basic supplies is released from a variety of altitudes to fly to isolated unit
Provide close support to friendly ground forces	Fighter-released UAVs use GPS/3-D maps/lasers to fly to and identify adversary's position
	UAV fires multiple grenade-sized explosives through window

For these capabilities to be realized, several areas will require more focused attention:

- Air-launched offboard sensors

- Limited-effects munitions and associated platforms

- Non-imaging sensors for ground networks (particularly weapon-detection and explosives-detection technologies)

- Three-dimensional mapping and databases

- Sensor fusion

- Joint command and control of aerospace and ground forces.

Budgetary realities and current modernization priorities mean that funds available for enhancing USAF urban capabilities are limited. For that reason, we recommend that the USAF continue modest research in this area to identify the most-promising and versatile technologies. Additional research and testing will have to be done before there is sufficient data on performance and cost for the USAF leadership to make informed decisions on whether to field systems such as those discussed in this report.

For the near term, we recommend that one of the USAF major commands or a battle laboratory be given responsibility to do additional research and development of these systems. To make the most of limited R&D funds, USAF laboratories should seek to partner with the Defense Advanced Research Projects Agency (DARPA) and other interested parties—perhaps under the auspices of an Advanced Concept Technology Demonstration (ACTD) program—to build and test prototypes of the more-promising systems. U.S. allies are likely to be important players as well; they have developed a variety of nonlethal, countersniper, and other systems that apply in urban operations.

Ultimately, urban operations are a joint problem. Theater commanders, the joint staff, and DoD will have to determine which mix of capabilities offers the most robust force for urban operations. Specific sensor and weapon choices will have to be made on the basis of some combination of coverage rate, resolution, versatility, responsiveness, cost, proportional/adjustable effects, and ease of delivery. Urban-specific measures of effectiveness (MOEs) may be needed to evaluate options for accomplishing the various tasks. As promising technologies are identified, realistic field testing, simulation, modeling, and red-teaming[2] will be necessary to determine which, if any, of them are sufficiently robust under real operational conditions to justify fielding.

Some of the possibilities discussed in this report, such as urban pattern analysis[3] and the fusion of aerospace-ground sensor inputs, belong in a joint fusion or command center. Therefore, these capabilities should be developed under joint initiatives. Indeed,

---

[2]A *red team* seeks to identify clever countermeasures that adversaries might develop to defeat U.S. systems or concepts.

[3]*Urban pattern analysis* involves collecting data on urban activities (e.g., movement of people or vehicles) in order to detect anomalies that might be associated with adversary activity.

some of the most difficult issues are related to joint command and control of urban operations. For example, coordinating joint fires to prevent friendly forces from firing on one another will become a bigger problem if the number of standoff weapons used in urban operations are increased. If significant numbers of friendly forces are on the ground, will all urban air strikes be considered close air support (CAS)? Or will aerospace forces operate more autonomously in some parts of the city? These are just a few of the many issues that need to be resolved before highly integrated urban aerospace-ground operations become feasible.

It would be unfortunate, however, if excessive concern about budgetary constraints, combined with somewhat outdated views of the limitations of aerospace power, prevented promising new capabilities from being fielded. Ironically, airmen are often as likely as infantrymen to narrowly define the settings in which aerospace forces can contribute. A more expansive vision of aerospace power would see the urban canyons of the world as part of the continuum of the vertical dimension that runs from the ground to orbital altitudes and would embrace nontraditional systems—such as air-dropped UAVs—as simply another tool in the airman's kit bag. The USAF excelled during the twentieth century at going higher, faster, farther. To meet the challenges of the early twenty-first century, the USAF may also need to exploit unmanned and robotic systems so that it can go lower, slower, and closer against unconventional threats to U.S. interests.

ACKNOWLEDGMENTS

The authors would like to thank the following individuals for their contributions to this study.

Lt Gen Norton Schwartz, then–Director of Strategic Planning, Headquarters U.S. Air Force (USAF), was the study sponsor. We very much appreciate his interest, enthusiasm, and helpful guidance as the project evolved. Col Robert Stephan and Maj Jeff Newell were the study action officers. They were extremely helpful on both the substance of the research and administrative support. Bob's tireless efforts to get this work to key audiences and Jeff's assistance in setting up a visit to the 16th Special Operations Wing (SOW) are especially appreciated.

We are grateful to GEN John Tilelli, U.S. Army (USA), Commander in Chief, U.S. Forces Korea; Gen Patrick Gamble, USAF, Commander, Pacific Air Forces; MG William Lennox, USA, Director of Operations, U.S. Forces Korea; and Maj Gen Steven Polk, USAF, then–Director of Operations, Pacific Air Forces, for sharing their observations and thoughts on urban operations during a project trip to the Pacific theater in June 1999.

At Air Force Special Operations Command, we want to thank Capt Tyler Sheppard, who helped arrange our visit to the 16th SOW. Within the wing, we want to thank Lt Col Mark Transue, Commander of the 16th Special Operations Squadron (SOS), for hosting our visit. Lt Col Greg McMillan, Director of Operations, 16th SOS, and Maj Greg Jerrell, Assistant Director of Operations, were extremely generous with their time, spending many hours with our team discussing AC-130 operations in the urban setting. Thanks are due to

Capt Smith and the crew of Spectre 62 for taking a project team member along on a 4-hour AC-130 training mission that included a live-fire exercise, as well as a dry-fire urban exercise. Maj Jerrell also kindly acted as escort during this flight. We also want to thank officers and NCOs from both the 16th and 20th Special Operations Squadrons who spoke with us about their real-world flying experiences in Panama, Somalia, and Liberia. Finally, MSgt Tim Wilkinson and TSgt Jeff Bray, veterans of the most ferocious small-unit action the United States has seen since the Vietnam War, shared their insights about "Bloody Sunday" in Mogadishu.

Others in the USAF who provided helpful comments include Lt Col Hank Andrews, Lt Col Mike Condray, Lt Col Phil Smith, and Maj John Hicks. Maj Jeffrey Walker, HQ USAF/JAG (Judge Advocate General), kindly arranged a very informative meeting with other JAG officers to discuss law-of-armed-conflict issues in the urban setting.

We also would like to thank the following individuals for their assistance in providing documentary materials relating to aerospace power's contribution to past urban operations: Dr. Wayne Thompson and Ms. Yvonne Kinkaid (Air Force History Support Office); Mrs. Essie Roberts and MSgt David A. Byrd (Air Force Historical Research Agency); Drs. Robert K. Wright and Richard Hunt (Army Center for Military History); and SFC Leo Daugherty and Mr. David Crist (Marine Corps Historical Center).

Professor Robert Ellefsen of San Jose State University kindly shared his research on urban terrain zones and provided access to his urban terrain database. Dr. Mark Shulman and Bob Haffa provided very helpful comments on parts or all of the draft report.

We are also indebted to the commanding officer of the Israeli Air Force (IAF), Major General Eitan Ben-Eliahu, for his support of a project delegation visit to Israel in early October 1999 in connection with this research and, in particular, to Brigadier General Shlomo Mashiah, head of the IAF Air Division for Helicopters, for sharing his insights on recent IAF experience with air operations against Hezbollah forces in and around the urban environment of southern Lebanon. Col Bill Clark, defense and air attaché and Lt Cols Joe Nichols and Bart Wohl, assistant air attachés, U.S. Embassy, Tel Aviv, kindly assisted with the administrative arrangements for this visit.

For their help in providing a critical sounding board for our ideas during the earliest phase of this study's formulation, we also wish to acknowledge Terry Stinson, Chairman and CEO; Dick Spivey, Director of Tiltrotor Business Development; and numerous other executives and senior analysts at Bell Helicopter Textron, Inc.

At RAND, we thank Stephen Hosmer, who acted as a senior adviser to the project. Although other commitments kept him from working formally on the project, Stephen participated in project brainstorming sessions, attended meetings with the USAF, and provided detailed comments on our briefing and written products. Other colleagues who helped improve our briefings and report include Dan Byman, Sean Edwards, Scott Gerwehr, Russell Glenn, John Gordon, Glenn Kent, Tom McNaugher, Rich Moore, Dick Neu, David Ochmanek, Don Palmer, Bruce Pirnie, William Rosenau, and David Shlapak.

Jennifer Casey, the Washington Office reference librarian, located data on land-use patterns, urban demographics, and a host of related subjects, and also discovered urban photos for wargaming exercises. Sandra Petitjean converted our PowerPoint graphics to EPS format. Dionne Sanders and Joanna Alberdeston prepared the manuscript. Their enthusiasm, professionalism, and responsiveness are greatly appreciated.

We acknowledge Randy Steeb and Barry Watts, our technical reviewers, for their thorough and insightful reviews, which greatly strengthened the report.

Finally, we thank Marian Branch, our tireless editor, for clarifying, unifying, and sharpening our prose. Her organizational suggestions also greatly improved the report.

ABBREVIATIONS AND ACRONYMS

AAA	Anti-aircraft artillery
ACN	Airborne Communication Node
ACTD	Advanced Concept Technology Demonstration
AFB	Air Force Base
AFP	Air Force Pamphlet
AGL	Above ground level
AGM	Air-to-ground missile
APC	Armored personnel carrier
ARL	Army Research Laboratory
ARVN	Army of the Republic of Vietnam
ATACMS	Army Tactical Missile System
ATGM	Antitank guided missile
ATO	Air Tasking Order
AWADS	Adverse Weather Aerial Delivery System
BDA	Battlefield damage assessment
C2	Command and control
C3I	Command, control, communications, and intelligence
C3ISR	Command, control, communications, intelligence, surveillance, and reconnaissance
C4ISR	Command, control, communications, computers, intelligence, surveillance, and reconnaissance
CALCM	Conventional Air-Launched Cruise Missile
CARP	Computer Aerial Release Point
CAS	Close air support
CCD	Charge-coupled device

CCT	Combat Control Team
CD	Aerodynamic drag coefficient
CDS	Container Delivery System
CENTAF	Central Air Force
CENTCOM	Central Command
CINC	Commander in chief
CINCUSFK	Commander in chief, U.S. Forces in Korea
CL	Aerodynamic lift coefficient
COMAIRSOUTH	Commander of NATO's air forces in the Southern Region
CS	Most widely used riot-control agent made from ortho-chlorobenzylidene malononitrile
CSAR	Combat Search and Rescue
DARPA	Defense Advanced Research Projects Agency
DEM	Digital Elevation Model
DoD	Department of Defense (U.S.)
DTED	Digital terrain elevation data
EFOG-M	Enhanced Fiber-Optic Guided Missile
EO	Electro-optical
EPRLRS	Enhanced Position Location Reporting System
FAA	Federal Aviation Administration
FAC	Forward air controller
FCC	Federal Communications Commission
FIBUA	Fighting in Built-Up Areas
FPA	Focal plane array
GIS	Geospatial information system
GPO	Government Printing Office (U.S.)
GPS	Global Positioning System
GRADS	Ground Radar Aerial Delivery System
HAETC	Hughes Advanced Electromagnetic Technology Center
HHSAR	Hand Held Synthetic Aperture Radar
HPM	High-powered microwave
IAF	Israeli Air Force
ID	Identification
IDF	Israeli Defense Force
INS	Inertial Navigation System

InSAR	Interferometric synthetic aperture radar
IR	Infrared
IRA	Irish Republican Army
IRCM	Infrared countermeasure
ISR	Intelligence, surveillance, and reconnaissance
IV	Intravenous fluid
JAG	Judge Advocate General
JDAM	Joint Direct-Attack Munition
JFACC	Joint Force Air Component Commander
JSTARS	Joint Surveillance and Target Attack Radar System
JTIDS	Joint Tactical Information Distribution System
KTO	Kuwaiti Theater of Operations
LAAS	Local Area Augmentation System
LANTIRN	Low Altitude Navigation and Targeting Infrared for Night
LGB	Laser-guided bomb
LOC	Line of communication
LOCAAS	Low-Cost Autonomous Attack System
LOS	Line of sight
LPI	Low probability of intercept
LWIR	Long-wave infrared
MAAF	Mediterranean Allied Air Forces
MANPADS	Manportable air defense system
MAV	Micro–aerial vehicle
MAWTS	Marine Aviation Weapons and Tactics Squadron
MEU	Marine Expeditionary Unit
MEW	Microwave Early Warning
MOBA	Military Operations in Built Up Areas
MOE	Measure of effectiveness
MOLA	Mars Orbiter Laser Altimeter
MOOTW	Military operations other than war
MOUT	Military Operation on Urbanized Terrain
MSI	Multispectral image (processing)
MSSI	Multispectral Solutions, Inc.
MTI	Moving Target Indicator
MWIR	Medium-wave infrared

NASA	National Aeronautics and Space Administration
NATO	North Atlantic Treaty Organization
NCO	Noncommissioned officer
Nd:YAG	Neodymium yttrium aluminum garnet
NEO	Noncombatant evacuation operation
NGO	Nongovernmental organizations
NIIRS	National Imagery Interpretability Rating Scale
NIMA	National Imagery and Mapping Agency
NKVD	World War II–era Soviet intelligence and internal security organization
NRL	Naval Research Laboratory
PDF	Panamanian Defense Forces
PGM	Precision-guided munition
PLO	Palestine Liberation Organization
PLRS	Position Location Reporting System (AN/TSQ-129)
POW	Prisoner of war
PPM	Pulse-position modulation
R&D	Research and development
RAF	Royal Air Force (United Kingdom)
REMBASS	Remotely Monitored Battlefield Sensor System
QRF	Quick-reaction force
RF	Radio frequency
ROE	Rules of engagement
ROK	Republic of Korea
RPG	Rocket-propelled grenade
SAM	Surface-to-air missile
SAR	Synthetic aperture radar
SEAD	Suppression of enemy air defense
SEALs	Sea, Air, Land (Navy Special Forces)
SINCGARS	Single-Channel Ground and Airborne Radio System
SOAR	Special Operations Aviation Regiment
SOF	Special operations force
SOS	Special Operations Squadron
SOW	Special Operations Wing
SPIN	Special Instructions

STO	Sensor Technology Office
TAC	Tactical Air Command
TACC	Tactical Air Control Center
TDOA	Time-difference-of-arrival
TOW	Tube-launched, optically tracked wire-guided (missile)
UAV	Unmanned aerial vehicle
UGS	Unattended ground sensor
UHTS	Ultra-high throughput screening
UN	United Nations
UNPROFOR	UN Protection Force
USA	United States Army
USAF	United States Air Force
USEUCOM	U.S. Command in Europe
USSTRICOM	U.S. Strike Command
UTZ	Urban Terrain Zone
UWB	Ultra-wideband
UWG	Urban Working Group
VTOL	Vertical takeoff and landing
WMD	Weapon of mass destruction
WWII	World War II

Chapter One

INTRODUCTION

BACKGROUND

For several millennia in Asia, Europe, the Middle East, and the Americas, the city has played a unique role as a center of economic, political, and cultural activity. As well, the city has long played a military role. Built as fortified settlements, many early cities have been used as anchors for defensive lines.[1] Typically astride major trade routes, cities—particularly national capitals—have often been the key to the physical and psychological vitality of a nation. Nations and armies have had varied motives in attacking cities. Annexation, tribute, destruction of the enemy's political center of gravity, denial of industrial or other economic resources, seizure of transportation hubs, defeat of enemy military forces, and the creation of refugees—all have been goals of attackers at various times and places.

The fall of major cities in war has typically been associated with final defeat. Thus, it is no surprise that battles for cities have been central to both civil and international conflicts from the Peloponnesian War to Bosnia. They also have been very common: Urban battles number in the thousands.[2] Progress in wars has often been measured with

---

[1]For a discussion of the role of cities in past wars, see G. J. Ashworth, *The City and War*, London: Routledge, 1991.

[2]A short list of twentieth-century urban battles includes Madrid (1936–1937), Stalingrad (1942), Bastogne (1944), Caen (1944), Berlin (1945), Manila (1945), Seoul (1950), Budapest (1956), Algiers (1957), Prague (1968), Hue (1968), Saigon (1968 and 1975), Jerusalem (1948 and 1967), Port Suez (1973), Khorramshar (1980), Beirut (1982), Panama City (1989), Kabul (1989–1991), Khafji (1991), Mogadishu (1993), and Grozny

respect to these cities. For example, during World War II (WWII), U.S. newspapers and newsreels routinely ran headlines such as "Allies Capture Caen" or "Allies 30 miles from Paris." Although rivers and national borders were also used as milestones, the capture of key cities was viewed by both national leaders and the public as the true measure of success. In short, unlike any other terrain feature on a map, cities have symbolic and practical significance.

Despite the importance of cities and extensive U.S. experience with urban combat in WWII and, to a lesser extent, in Korea and Vietnam, urban military operations received little attention during the Cold War. However, since the end of the Cold War, the U.S. defense community has grown increasingly interested in urban combat.[3]

Since the end of the Cold War, U.S. forces have been involved in a number of operations (peacekeeping, humanitarian relief, and non-combatant evacuations) that have taken place in urban settings. Peace operations in Somalia, especially the deaths of 18 U.S. servicemen and the wounding of almost 100 others on October 3, 1993, profoundly influenced the American public's perceptions of modern urban combat in the developing world.[4] For military professionals, Somalia also was a painful reminder that the technological and operational dominance the United States experienced on the conventional battlefield during Desert Storm did not necessarily carry over into urban peacekeeping. For infantrymen in particular, the fierce fighting of "Bloody Sunday"—the most intense light infantry engagements since the Vietnam War—brought home the relevance of urban combat, its nastiness, and the need to develop concepts and tactics better suited to this unique environment.[5]

(1996). There were probably hundreds of urban battles in small towns and other built-up areas in World War II alone.

[3]For example, the December 1997 National Defense Panel report (p. 21) highlighted urban operations as increasingly likely in future conflicts.

[4]For a riveting account of this grim episode, see the now-classic narrative by Mark Bowden, *Black Hawk Down: A Story of Modern War*, New York: Atlantic Monthly Press, 1999.

[5]For a thoughtful assessment of these challenges, see Russell W. Glenn, *Combat in Hell: A Consideration of Constrained Urban Warfare*, Santa Monica, Calif.: RAND, MR-780-A/DARPA, 1996.

HOW LIKELY ARE URBAN OPERATIONS?

Are recent urban operations in Panama City, Khafji, Mogadishu, Port au Prince, Grozny, and Sarajevo an anomaly or a harbinger of the future?

According to a variety of hypotheses that have been posited,[6] urban military operations will become more frequent.

By all accounts a fundamental demographic transition does seem to be occurring in the developing world. The world's urban population is growing four times as fast as its rural population, and 150,000 people are added to the urban population of developing countries every day. By 2025, two-thirds of the earth's population is projected to live in urban areas, and 90 percent of the growth will be in the developing world.[7] Together, population growth and migration are leading to an urban world and changing the fundamental character of many previously agrarian societies.

Some observers believe that, as populations shift from rural to urban areas, the focus of existing conflicts—whether tribal, ethnic, religious, or ideological—will shift to urban areas also. For example, insurgents in El Salvador, Peru, Angola, Liberia, Sierra Leone, Afghanistan, and Egypt all shifted their focus to cities over the past two decades. In some cities, the slums have become urban sanctuaries for insurgents, much as isolated rural settings were during the insurgencies of the 1960s.[8]

Although some conflicts with rural roots have shifted to urban areas, it should not be assumed that migrants will necessarily carry their rural animosities into the city. Several hundred years of rural-to-urban migrations have shown that cities change people's attitudes and be-

---

[6]See, for example, Russell W. Glenn, *Marching Under Darkening Skies: The American Military and the Impending Urban Operations Threat*, Santa Monica, Calif.: RAND, MR-1007-A, 1998.

[7]*World Resources 1998–1999: A Guide to the Global Environment*, Oxford: Oxford University Press, 1998, p. 146; *World Resources 1996–1997: A Guide to the Global Environment*, Oxford: Oxford University Press, 1996, p. 4.

[8]Jennifer Morrison Taw and Bruce Hoffman, *The Urbanization of Insurgency: The Potential Challenge to U.S. Army Operations*, Santa Monica, Calif.: RAND, MR-398-A, 1994, pp. 12–15.

haviors. Birthrates drop, political allegiances shift, economic interests are often radically altered. It is difficult to assess the net effect of these changes in the abstract. In some cases, migration may remove sources of conflict; in others, it may produce the seeds of new urban-derived violence.

For example, some believe that the marginal living conditions of some urban areas in the developing world, particularly in the unsanctioned shantytowns that surround them, will lead to unrest.[9] Although these conditions are often appalling from a Western point of view, they have to be assumed to be an improvement over rural life since most urban migration is voluntary. On the one hand, if migrants see their lives gradually improving and believe that the shantytown life is a temporary step on the way to a better existence in more-permanent urban dwellings, they are unlikely to be interested in political violence. On the other hand, rising expectations, if not met by improved conditions, could lead to resentment and provide a pool of possible recruits for criminal, terrorist, or insurgent groups. That said, we must recognize that the causes of violence—both criminal and political—are complex. No single variable (e.g., level of poverty) is a reliable indicator of a future propensity to violence.

Yet another argument for an increase in urban conflict sees urbanization as denying the open space to conduct traditional maneuver warfare. For example, Rosenau argues that "the amount of open space is decreasing, thus increasing the odds that land forces will have to fight in urban areas."[10] This certainly is true in two places in which the United States has historic interests: Central Europe and the Korean Peninsula. It would be very difficult to fight a major war in either location without a substantial urban component.

That said, land-use patterns by themselves are not sufficient to argue for an increase in urban conflict globally, because the amount of urbanized terrain is still small compared with open space—forest,

---

[9]For darker views of these trends, see Eugene Linden, "The Exploding Cities of the Developing World," *Foreign Affairs,* January–February 1996, pp. 52–65; and William G. Rosenau, "Every Room Is a New Battle: The Lessons of Modern Urban Warfare," *Studies in Conflict and Terrorism,* Vol. 20, 1997, p. 374.

[10]Rosenau, 1997, p. 372.

farm, mountain, or desert. For example, in 1994, 37 percent of the earth's land surface was farmland or pastureland.[11] Data for the percentage of the earth's land surface that is undeveloped are hard to come by; however, population-density and land-use graphics in the 1998 *World Resources Guide* suggest that well over 50 percent, and probably closer to 75 percent, of the earth's land surface is nonurban: farmland, pastureland, grassland, forest, desert, or tundra.[12]

Another argument is that future U.S. adversaries will not want to fight wars in open spaces (at least not against the United States) and will actively avoid confronting the United States in places and under conditions in which its advanced sensors and weapons are most effective. Rather, among other clever strategies, they will take advantage of the land-use patterns that Rosenau identified and seek out built-up areas to counter U.S. technological dominance of the conventional battlefield. Their ability to do so will vary with their objectives and a host of military and political factors. However, the United States should expect future adversaries to increasingly embrace such strategies.

Perhaps the most compelling argument for future urban operations is also the most simple: Many U.S. objectives cannot be achieved without controlling key cities (or parts of them) for some period of time. For a special operation, this control might last for a matter of minutes or hours; for a noncombatant evacuation operation (NEO), it might last for days; for peace operations, it might last for years. We saw this in recent operations in Somalia, Panama, Haiti, Kuwait, and Bosnia, where U.S. forces were assigned cities as major operational objectives. Noncombatant evacuations—such as those that occurred in Somalia (1991), Liberia (1996), the Central African Republic (1996), Republic of Congo (1997), and Sierra Leone (1997)—are invariably centered on the capital or other major cities.

Cities were also core objectives in other operations in the 1990s. In Kuwait, the capture of the country's capital city was an essential part of the larger objective of liberating Kuwait. In Panama, removing the national leadership and disarming Panamanian Defense Forces

---

[11] *World Resources 1998–1999*, p. 298.

[12] *World Resources 1998–1999*, pp. 221–222.

(PDF) were the core objectives, requiring the capture of PDF facilities in and around Panama City. In Haiti, control of Port-au-Prince, Cap-Haïtien, and other cities was necessary to ensure that General Raoul Cedras and the Haitian military would step down as they had agreed, to prevent violence and to allow President Jean-Bertrand Aristide to be restored to power. In Somalia, Mogadishu was the only place with the infrastructure (a port and an airfield) necessary to support the humanitarian intervention. It also was the scene of factional fighting and the headquarters for the Somali National Alliance and other factions. Similarly in Bosnia, it would have been impossible to enforce a peace without controlling the cities.

THE NATURE OF URBAN MILITARY OPERATIONS

Urban combat, Fighting in Built-Up Areas (FIBUA), Military Operations in Built Up Areas (MOBA), Military Operations on Urbanized Terrain (MOUT), operations in complex terrain, and urban military operations are all terms used to describe the subject of interest in this report. We use *urban military operations* in this report to describe any military activity from humanitarian relief to conventional combat that occurs in built-up areas. For our purposes, any area in which man-made structures are the dominant terrain feature—whether a large city, small town, or village—is considered *urban*. From a tactical perspective, any area sufficiently built up that it channels the movement of forces, restricts fields of fire, extends infantry combat vertically above and below the surface of the earth, and provides defenders a multiplicity of "natural" defensive strong points, concealment, and the potential for unobserved movement through buildings, is *urban*.

A city, however, is more than just a physical environment. It is a political, economic, social, and psychological environment as well. The physical landscape of the city, both natural and man-made, is the shell on which and within which a vast and multilayered living organism—the urban population—lives. The interplay of private and public activities—in homes, businesses, schools, marketplaces, and government—that constitutes urban life adds hundreds of degrees of complexity beyond those presented by the urban physical environment. In short, unconstrained urban combat in the abandoned shell of an empty city, although enormously difficult from a tactical perspective, is straightforward compared with military operations in a city whose civil lifeblood still pulses.

From Rubble to Routine

During the Cold War, most defense professionals understood urban operations to mean house-to-house fighting between conventional ground forces seeking to hold or take a city in the context of a major war such as World War II. Fighting of this intensity usually turned city landscapes to rubble and disrupted the routine economic and social activities of cities, typically causing mass evacuations and refugee flows. Thus, as urban battles wore on, the civilian population and its associated activities became less and less a factor. The all-out nature of WWII also meant that commanders were focused on defeating enemy forces whatever the cost. To constrain operations because of concerns about civilians would have risked defeat. The urban-operations challenge was primarily to find solutions to the tactical, logistics, engineering, and command exigencies of operating in this unique physical environment.[13] Lesser conflicts that started in or spread to urban settings generally have been considered to be a subset of low-intensity conflict (what we now call MOOTW) typically not discussed in analytical or doctrinal writings on urban combat.[14]

Recent writings on urban operations, in contrast, have emphasized the challenges of lesser conflicts in urban terrain.[15] The typical scenario is one in which U.S. and allied forces are attempting to enforce a peace agreement or defeat an insurgency in an urban setting. This

---

[13]Admittedly, there are likely to be some situations in major wars where the civilian population would be a significant factor. For example, if Seoul were captured early in a future Korean conflict, the fate of Seoul's civilian population would likely weigh heavily on allied planners as they developed options to liberate the city. Depending on North Korean treatment of the civilian populace, Seoul might even be bypassed initially in a counteroffensive to spare the city and population.

[14]For example, the 1979 version of Army Field Manual 90-10, *Military Operations on Urbanized Terrain (MOUT)*, makes no mention of the civilian population. This manual is clearly focused on the tactical challenges of defeating conventional enemy forces in a city that has been largely abandoned by its civilian population, although it never states this explicitly. See also Russell W. Glenn, "*. . .We Band of Brothers,*" Santa Monica, Calif.: RAND, DB-270-A, 1999.

[15]See, for example, Ralph Peters, "Our Soldiers, Their Cities," *Parameters,* Spring 1996, pp. 43–49.

scenario—reminiscent of U.S. operations in Mogadishu—envisions an environment in which routine civilian functions and small-unit combat are juxtaposed; in which adversary forces have at least the tacit support of some of the population; in which adversary forces can exploit the physical and human landscape of the city for concealment, support, intelligence, mobility, and tactical advantage; in which strict rules of engagement (ROE) limit the use of heavy weapons; and in which U.S. military actions are monitored closely by international television and press.

These lesser conflicts have received more attention in the military literature both because they are more difficult to plan for than all-out war and because many believe such conflicts will become increasingly frequent, which has led to some definitional confusion in the broader defense community.

The urban-operations writings are paralleled by a more bureaucratic debate in Washington, and some services have seized upon urban operations as a mission they wish to claim as theirs exclusively.

Redefining Warfare, Not Service Roles

Some authors and senior U.S. Marine officers use *urban operations* increasingly as a shorthand for what they believe will be the most prevalent form of conflict in the future—civil wars, insurgencies, and transnational terrorism in the world's cities—rather than its traditional meaning of all combat in built-up areas.[16] These observers believe that subnational groups will increasingly attack U.S. interests through urban-based operations that do not present targets for the decisive application of large-caliber firepower, including precision weapons delivered by aircraft. In essence, the subject these authors and service representatives are seeking to capture is the future of warfare, not how best to conduct traditional military operations in urban settings.

Airmen have responded to this new view of urban military operations in a variety of ways. Some have noted that aerospace power has

---

[16]The Marines have focused on these complex, highly constrained contingencies in their Urban Warrior exercises.

played an important role in most urban battles the United States has fought, either through interdiction of enemy forces or by providing surveillance, reconnaissance, aerial resupply, and/or close support to friendly ground forces. Other airmen have simply dismissed urban-based lesser conflicts as secondary in importance to major wars. Still others have introduced their own redefinition, arguing that precision strikes in Baghdad or Bosnia during Operations Desert Storm and Deliberate Force constituted urban combat.

Rather than embrace any one of these perspectives, we suggest that defense planners cannot afford to focus on one type of urban conflict to the exclusion of others. Defense planners need to identify those capabilities that the United States will need for diverse urban operations that will vary on at least the eight dimensions listed in Figure 1.1.

As we think through the many potential permutations, it becomes clear that the kinds of capabilities the United States needs for urban military operations are going to vary greatly, depending on the par-

RAND *MR1187-1.1*

> Nature of opponent (e.g., nation, subnational group)
>
> Type of adversary forces (e.g., regular, irregular, heavy, light)
>
> Rules of engagement (e.g., restrictive to permissive)
>
> Physical environment (e.g., high-rise or low-rise, compact or sprawl, permanent or shantytown, town or city)
>
> Social environment (e.g., degree of popular support for adversary, city populated or evacuated)
>
> U.S. objectives (e.g., enforce peace agreement, liberate friendly city)
>
> Adversary's objectives (e.g., capture territory, overthrow government)
>
> Adversary's strategy (e.g., city hugging, urban insurgency)

NOTE: *City hugging* occurs when conventional forces use urban terrain to hinder the defender's attempts to detect and attack them with standoff sensors and weapons.

Figure 1.1—Analytic Dimensions of Urban Military Operations

ticular situation. For example, at least some of the tasks[17] the U.S. military would be called upon to accomplish in an urban battle during a major theater war can be expected to be different from those associated with an urban counterinsurgency operation, urban peacekeeping, or opposed NEO in an urban setting. Even when the tasks are the same (e.g., detect, identify, and attack adversary forces), the means used to accomplish them will vary according to the rules of engagement, social environment, and so on. Consequently, the relative contribution of air, land, sea, and space forces will also vary with the specifics of a particular scenario.

It would be risky to posture U.S. forces for only one of the many potential scenarios or to rely exclusively on any one type of force element for the diverse challenges of urban military operations. The smoke from the interservice competition for resources should not obscure the fact that each service has roles to play in urban operations.[18] Indeed, one huge advantage that the United States enjoys over many potential foes is the ability to integrate and orchestrate joint forces to accomplish key operational tasks. Urban-operations analysis should cover the spectrum of possible urban-conflict situations, identifying operational tasks associated with each one and deriving weapons, training, tactics, and organizational requirements from these tasks.

PURPOSE

The objective of this report is to help the USAF and others in the defense community better understand the role of aerospace forces in future urban military operations. Because urban stabilization operations—peace operations, counterinsurgency, or humanitarian

---

[17]One challenge for future researchers in this area is to develop task lists for various missions in an urban environment.

[18]The role of the Navy in urban operations tends to get less attention than it deserves. With most of the earth's population within 200 miles of a coastline, naval forces have participated in many urban operations, particularly noncombatant evacuations. Naval contributions to urban operations have included or could include transportation, surveillance, airlift, and fire support. Urban operations in port cities might also require use of naval SOF patrol craft, mine warfare assets, and smaller surface combatants. Navy SEAL teams are particularly well suited to operate in such environments.

aid—are likely to remain the most common, the emphasis of the report is on improving USAF capabilities to tackle these problems.

ORGANIZATION

Recognizing the difficulty of predicting future military challenges and seeing a role for aerospace forces in urban operations across the spectrum of conflict, this report addresses both conventional and unconventional challenges. Chapter Two explores how aerospace forces can be used to deter or prevent conventional attacks on urban areas. Chapter Three discusses the unique legal and political constraints on urban military operations, from classic strategic air campaigns to peace operations.

The remainder of the report focuses on enhancing the contribution of air and space components in lesser urban conflicts. Chapter Four presents data on urban geospatial forms and analyzes how the urban physical environment constrains air operations. Chapter Five presents new concepts of operation to accomplish key military tasks in light of the challenges and opportunities identified in the previous chapters. Chapter Six identifies key technology areas in which investments will be necessary to achieve the capabilities envisioned in Chapter Five. Chapter Seven presents the conclusions and recommendations. Appendix A shows how trigonometric calculations in Chapter Four were done. Appendix B discusses microwave recharging of UAVs. Appendix C provides additional details on countersniper technologies. Appendix D presents an analysis of lessons learned from previous urban air operations.

Chapter Two

USING AEROSPACE POWER TO PREVENT THE URBAN FIGHT

INTRODUCTION

A common tendency of recent discussions of urban military operations has been to portray urban combat as the most likely form of global strife in the decade ahead. Former Marine Corps commandant General Charles Krulak set the tone for such portrayals when he declared in 1996 that "the future may well not be 'Son of Desert Storm,' but rather 'Stepchild of Somalia and Chechnya.'"[1] Since the U.S. Army found itself caught so unprepared for the urban showdown it experienced three years earlier in Somalia, in which 18 soldiers were killed in vicious firefights in the streets of Mogadishu, other specialists in urban warfare have aired similar views. In summarizing what he saw as the implications of this trend, a former Army officer who has written extensively on urban operations concluded that "in the future, the term 'urban warfare' will be a redundancy."[2] Similarly, the highly regarded urban operations manual recently developed by Marine Aviation Weapons and Tactics Squadron (MAWTS) 1 accepted urban fighting as the look of tomorrow in its lead sentence:

---

[1]Gen. Charles C. Krulak, USMC, "The United States Marine Corps in the 21st Century," *RUSI Journal*, August 1996, p. 25.

[2]LTC Ralph Peters, USA, "The Future of Armored Warfare," *Parameters*, Autumn 1997, p. 56.

"Most future conflicts will involve Military Operations on Urban Terrain (MOUT)."[3]

Pervading such characterizations has been a tacit presumption that, because U.S. forces in all services enjoy increasingly pronounced air, space, and standoff dominance, tomorrow's adversaries will be driven to make the United States a victim of its own success by choosing to fight from embattled cities, where American military advantages will have the least leverage for getting close to and deterring or killing an adversary. Even more striking has been their tendency to treat urban operations in isolation from the broader theater campaign. For example, a concept of operations proposed by the Marine Corps Combat Development Command to accommodate the accelerating rate of worldwide urbanization called for beginning with "understanding the urban environment" and proceeding from there to developing tactics and procedures; exploring new technologies geared to the needs of maneuver warfare in urban terrain; enhancing the ability of U.S. forces to train in urban settings; weighing better organizational approaches; and integrating all these activities in pursuit of a viable urban operations repertoire.[4]

As well, such characterizations have tended, almost as a matter of course, to treat urban operations as a problem to be dealt with primarily by ground forces. Even airmen who have devoted careful thought to the matter have appeared disposed to accept such a conclusion as self-evident. For example, in reporting the main lessons drawn from a 2-year urban CAS tactics development and evaluation exercise conducted by the 422nd Test and Evaluation Squadron at Nellis AFB, Nevada, a USAF weapons officer observed that "urban warfare . . . is intrinsically an infantry fight; door-to-door, street-to-street, and block-by-block. Traditional artillery or mortar support can be easily minimized by urban terrain, leaving the foot soldier to improvise, adapt, and overcome on his own. Rotary-wing fire support is vulnerable to attack from high buildings, and fixed-wing CAS is extremely difficult in the quagmire of densely packed city

---

[3]U.S. Marine Corps, *Aviation Combat Element: Military Operations on Urban Terrain Manual*, MCAS Yuma, Ariz.: MAWTS-1, Edition VII, August 1998, pp. 1-1, 1-11.

[4]U.S. Marine Corps, Combat Development Command, *A Concept for Future Military Operations on Urban Terrain*, Quantico, Va., July 25, 1997, pp. III-18–III-19.

streets and buildings." Accordingly, this airman concluded, "infantry will lead" in urban combat contingencies.[5] Similarly, a noted RAF aerospace power expert remarked in 1996 that "aircraft have traditionally not been well-suited to operating in densely populated urban environments."[6]

As most of these characterizations do, it is entirely reasonable to assume that in the post–Cold War era, modes of conflict will shift away from long-familiar patterns of the past and present new force-employment challenges to U.S. policymakers. In light of that prospect, it also is reasonable to assume that urban military situations with the potential to affect U.S. interests will increase in frequency. Yet, arguably, there has been a tendency among some to overextrapolate from recent events in Somalia, Bosnia, and Chechnya, which have involved urban combat as their defining features, to conclude that urban military operations will be part and parcel of future U.S. combat involvement. The commandant of the U.S. Army War College, Major General Robert Scales, recently spotlighted just such an eventuality when he complained that "there has been too quick a leap beyond the more conceptual aspects of war in urban terrain and into the weapons and tactics necessary to fight street-to-street and door-to-door."[7]

Most discussions of this subject tend to fixate on tactics that are essentially reactive,[8] to assume that aerospace power is too indiscriminate a weapon to be of great use to theater commanders confronted with urban challenges and that the sorts of precision air-delivered munitions that work so well against tanks and other targets in the open would be counterproductive in a more close-in urban situation. However, it does not follow from the proposition that urban challenges may be on the rise around the world that the United States is obliged to deal with them solely in a reactive way. As

---

[5]Maj Brooks Wright, "Urban Close Air Support: The Dilemma," *USAF Weapons Review,* Summer 1998, p. 17.

[6]Air Commodore Andrew G. B. Vallance, RAF, *The Air Weapon: Doctrines of Air Power Strategy and Operational Art,* New York: St. Martin's Press, 1996, p. 92.

[7]MG Robert H. Scales, Jr., USA, "The Indirect Approach: How U.S. Military Forces Can Avoid the Pitfalls of Future Urban Warfare," *Armed Forces Journal International,* October 1998, p. 68.

[8]For examples, see Krulak, 1996, and Peters, 1997.

General Scales has pointed out, urban contingencies may indeed be a harbinger of a future style of warfare that could affect U.S. interests.

The most attractive option available to U.S. decisionmakers and theater commanders in the face of such contingencies may well be to "preempt the enemy from using complex [i.e., urban] terrain in the first place"[9]—for an abundance of good reasons: To begin with, towns and cities make for a singularly forbidding combat venue. As anyone who has seen the popular movie *Saving Private Ryan* can well appreciate, fighting in urban spaces is exceptionally difficult, manpower-intensive, and bloody. Surveillance is hindered, because a high percentage of urban activities take place either indoors or underground, beyond the scrutiny of line-of-sight sensors. The likely presence of a civilian population further complicates operations and provides false targets for sensors. Indoor targets, the density of structures, and strict rules of engagement limit standoff options.

For these and other reasons, it is doubtful whether the United States will *ever* enjoy the comparative force-employment advantages in urban combat that it now enjoys for more open engagements against fixed targets and fielded forces. Instead, U.S. commanders will have little choice in such cases but to fight in less-than-ideal tactical circumstances. Because of these uncongenial facts, it has long been a given in U.S. military doctrine and practice that "commanders should avoid committing forces to the attack of urban areas unless the mission absolutely requires doing so."[10]

Beyond these reasons, the United States has an overarching policy reason to refrain from engaging in high-intensity urban combat: "We cannot destroy or significantly damage the infrastructure of a foreign urban center in pursuit of mission attainment and expect the population to remain friendly either to U.S. forces or those we support."[11] This cautionary note against winning a battle with the best of tomorrow's urban combat assets at the cost of high collateral damage is

---

[9]Scales, 1998, p. 68.

[10]U.S. Department of the Army, *Operations*, Washington, D.C.: Field Manual 100-5, May 1976, p. 81.

[11]Office of the Under Secretary of Defense for Acquisition and Technology, *Report of the Defense Science Board Task Force on Military Operations in Built-Up Areas,* Washington, D.C., November 1994, p. 8.

tantamount to the hospital situation in which the operation is a success but the patient dies—something to be avoided at all costs.

As later chapters of this report show, U.S. aerospace power already has the potential to do more by way of helping theater commanders mitigate the worst aspects of urban operations than it is usually given credit for. For example, aerospace power can help isolate cities through enforcement of no-fly or no-drive zones. Moreover, with selected technology improvements and new concepts of operations aimed at making the most of existing systems (e.g., AC-130 gunships, UAVs, and the Joint Surveillance and Target Attack Radar System [JSTARS]), U.S. aerospace power should be able to contribute even more toward making urban operations less onerous for friendly ground forces.[12] Because of their obligation to support the theater commander in chief (CINC), airmen have as their first challenge the timely and effective use of aerospace power to mitigate threats to other force elements in circumstances in which urban operations end up being unavoidable. Second, in seeking to make the most of what aerospace power has to offer, airmen must consider how best to use that power to preclude having to engage in urban combat in the first place. This chapter focuses on the second of these two challenges. It offers ideas on how U.S. aerospace power might be used to deter or stop attacks on friendly urban centers and briefly discusses the Israeli experience in Lebanon, as well as the U.S. experience in the Gulf War. Finally, the battle for Grozny, Chechnya, is used to illustrate how not to use aerospace power in an urban setting.

HOW AEROSPACE POWER CAN FORESTALL URBAN OPERATIONS

Aerospace power offers U.S. decisionmakers a unique opportunity for avoiding some of the uglier urban situations. Indeed, its potential

---

[12]In this context, as in all others in this report, *aerospace power* is a convenient shorthand expression that embraces not only air vehicles, sensors, and munitions, but also space-based intelligence, surveillance, and reconnaissance (ISR) and all the intangibles, such as training, tactics, command and control, and concepts of operations, that allow those hardware components to deliver desired operational effects. Aerospace power, moreover, is not limited to USAF equities but entails a medium of force employment in which *all* services have something important to contribute.

for keeping urban showdowns from getting out of hand is arguably aerospace power's single longest suit in the urban-warfare context. In a typical scenario, U.S. or allied forces are attempting to enforce a peace agreement or defeat an insurgency. Efforts to use aerospace power as a way of precluding urban combat in such scenarios will be driven, first, by a desire to avoid needless friendly casualties. They also will appeal to planners because of their ability to reduce the chances of civilian casualties, as occurred in Somalia during the infamous "Bloody Sunday" shootout in Mogadishu.

Anatomy of an Urban Military Operation

It is worth dissecting what happened in Mogadishu because it illustrates the complexity of contemporary urban military operations, particularly in making distinctions between combatants and noncombatants. Somali fighters, who wore no uniforms or distinctive clothing, hid behind mobs of unarmed men, women, and children as they moved toward the helicopter crash sites and as they fired on U.S. personnel. Although some of these mobs were composed of curious bystanders, the overwhelming majority were acting in support of the Somali fighters. Many times, when a fighter was shot and fell to the ground, an unarmed "noncombatant" in the crowd would pick up the weapon and begin firing at U.S. troops. Entire Somali families acted as gun teams (dad firing, mom propping up the weapon, and kids sitting on top); others helped as spotters or ammo carriers. In one famous incident, a U.S. Ranger refrained from firing at a woman with a baby until she pulled a weapon from behind the baby and began firing.[13] Under these conditions it is impossible to say with any certainty which of the casualties were "combatants" and which were "noncombatants." Although certainly some innocents (mainly children) were killed or wounded, the overwhelming majority either were actively engaged in hostilities against UN forces or deliberately put themselves in harm's way to watch the fighting. By any reasonable definition, the vast majority of the people on the streets around Task Force Ranger and the relief convoys were combatants.

---

[13]Bowden, 1999, p. 106.

This experience suggests that tidy distinctions between armed forces and civilians have little relevance in at least some of the environments in which U.S. forces find themselves operating, and it raises difficult questions about just what ROE and weapons are appropriate for situations where the civilian population of a city rises up against U.S. forces. Although nonlethal weapons may offer a solution, avoiding these situations altogether is more desirable.

Comparative Advantages of Aerospace Power

In pursuing such an approach, it makes sense to take a CINC's view of the problem from the top down, rather than *either* a going-in aerospace power perspective or a bottom-up ground perspective. Such a view offers a ready antidote against the danger of succumbing to the law of the instrument and predetermining the answer.[14] For another, it allows the best ways of accommodating various contingencies to be focused on, irrespective of the instrument, by recognizing that, in some cases, aerospace power may offer synergies with other force elements, if not a single-point solution, whereas, in other cases, it may not be even remotely the right tool. In contrast, taking a narrower approach makes the error of presuming that urban fighting has already become an accomplished fact. By the same token, taking an approach that asks, as its first question, what role aerospace power can play in accommodating urban challenges risks biasing the choice of options in favor of an aerospace power solution from the start.

First, aerospace power can rapidly project U.S. power and presence, including people and modest amounts of equipment, worldwide. In situations where enemy conventional forces threaten invasion, the likelihood of urban combat occurring will be inversely related to how quickly U.S. aerospace power can be brought to bear. Conventional urban combat is likely to occur in settings where there is not a nearby U.S. military presence. For such a presence to be realized, suitable nearby operating bases must be readily available.

---

[14]The *law of the instrument* stipulates that when all one has is a hammer, everything in sight looks like a nail in need of pounding.

Second, aerospace power can observe from the highest of high "ground" and, in so doing, expand the situational awareness of friendly forces at all levels. Air and space surveillance systems can aid in intelligence preparation of the battlefield when a developing contingency might threaten to degenerate into an urban confrontation. UAVs, for example, can make imagery available to tactical commanders and provide near-real-time updates as necessary. JSTARS and other airborne systems with Moving Target Indicator (MTI) modes can detect and track moving vehicles in the open, day or night, in most weather. Such an information edge can help maximize U.S. maneuver advantages by enabling the bypassing of defensive urban fortifications and allowing precision attack systems to find and take out deployed enemy forces in the open. It also can provide needed awareness and force protection for U.S. forces potentially deployed in harm's way. The result will be to reduce the friction and uncertainty that might otherwise beset a U.S. or allied ground commander.

U.S. air- and space-based ISR capabilities already offer greatly improved situational awareness for all command echelons in a joint operation. They cannot, at least not yet, address the legitimate concern voiced by some land warriors over finding and identifying a notional "enemy company in the basement of a built-up area" or "the 12 terrorists mixed in with that crowd in the village market."[15] However, they are more than adequate for supporting informed and confident force-committal decisions by a CINC against enemy formations on the move in the open. For all its continued limitations, such an advantage entails a major breakthrough in targeting capability and one that, in conjunction with precision attack systems, has made for a uniquely powerful force multiplier.

Third, aerospace power can deny the adversary a night sanctuary and, from standoff ranges, can destroy an adversary's fixed assets in his homeland, with virtual impunity. Other direct-action opportunities for aerospace power in preventing urban combat include cutting off an adversary's sources of resupply, creating and enforcing exclu-

---

[15]LtGen. Paul K. Van Riper, USMC (Ret.), quoted in *Clashes of Visions: Sizing and Shaping Our Forces in a Fiscally Constrained Environment*, Proceedings of CSIS-VII Symposium, October 29, 1997, Washington, D.C: Center for Strategic and International Studies, 1998, p. 38.

sion zones, and enforcing no-drive zones. Working with friendly ground forces, aerospace power can also help seal off parts of a city, physically isolating an adversary from the population and, by jamming or preempting commercial radio and television stations, psychologically isolating him from encouragement by the population. It can destroy or neutralize detected adversary armor before it gets into built-up areas, substantially reducing the threat to friendly ground forces. In addition, it can cue friendly special operations forces conducting urban-containment missions and protect the movement of people and material from base to base.

Fourth, aerospace power can deny entry to an adversary. This option is perhaps the most challenging task for aerospace power, since air and space sensors and air-delivered weapons cannot easily prevent clandestine units from entering cities. Yet, with timely warning and adequate presence, these sensors and weapons can deny potential adversaries the option of pursuing city-hugging strategies by eliminating their ability to move into a defender's cities, confronting friendly forces with a fait accompli.

The ability of aerospace power to do all of this with minimal friendly losses, thanks to stealth and advanced suppression of enemy air defense (SEAD) capabilities, makes such preclusion options particularly credible.

All in all, today's U.S. air and space capabilities enable constant pressure on an adversary to be maintained from a safe distance, increased kills per sortie, selective targeting with very little unintended damage, substantially reduced reaction time, and, at least potentially, the complete shutdown of an adversary's control of large mechanized forces.[16]

Preclusion Strategies

As for preclusion strategies themselves, theater CINCs can consider first using aerospace power to disincline adversaries from indulging

---

[16]For further development of these points, see Lt Gen George K. Muellner, USAF, "Technologies for Air Power in the 21st Century," paper presented at a conference on "Air Power and Space—Future Perspectives" sponsored by the Royal Air Force, Westminster, London, England, September 12–13, 1996.

in threats against friendly cities. For example, at the most benign end of the conflict spectrum, the timely resort to aerospace power can preempt urban fighting by providing early humanitarian assistance. That failing, CINCs can consider using aerospace power, along with early-entry ground forces, to foreclose enemy options by sealing off a town, blocking avenues of approach within cities, and otherwise making the urban environment nonpermissive before the situation degenerates into an urban nightmare. The challenge for the Joint Force Air Component Commander (JFACC) in such cases will be to determine what enables an adversary to engage in urban combat and then to take that ability away preemptively.

In this regard, the earlier-cited perspective offered by Army MG Scales (1998) would seem to be almost tailor-made for making the most of American air and space assets in heading off the worst kinds of urban warfare contingencies. By being patient and playing a waiting game, U.S. commanders can strive to run would-be urban opponents out of both time and latitude to control the local populace. Such use of the awareness and direct-action opportunities made possible by aerospace power could include augmenting host-nation forces, prepositioning equipment, and blocking avenues of approach to cities. Those options failing, aerospace power might still enable a CINC to keep urban combat at arm's length by avoiding a direct assault on adversary forces, establishing a cordon around a city, controlling the flow of information into the city (through electronic jamming and friendly psychological-warfare radio/TV broadcasts), creating sanctuaries to which the populace could escape, and enforcing safe passage for those who wished to escape. For such a preclusion strategy to work, General Scales cautions that planners would require reliable information on the extent of enemy popular support, on the population's willingness to endure suffering, on the city's ability to be self-sustaining, on the presence of defensible nearby sanctuaries, and on the coherence of occupying adversary forces, among numerous other things.[17] The main drawback of his suggested approach is that in many, if not most, urban contingencies that might arise to challenge U.S. interests, this option may not be available to U.S. commanders. For example, it might not be politically possible to play a waiting game if a friendly city was cap-

---

[17]Scales, 1998, p. 70.

tured by adversary forces, particularly if those forces were abusing the civilian population. Urban sieges are best applied against cities in the adversary's homeland or against smaller towns that had largely been evacuated.

Aerospace power can deter aggression or compel an end to aggression directed against cities by threatening strikes against known adversary centers of gravity, including targets in their homeland cities. In Operations Allied Force and Deliberate Force, and during the earlier 1991 Gulf War, U.S. aerospace power demonstrated an ability to identify key military, political, and economic targets and destroy them with great lethality and precision.

Aerospace power also can detect and destroy adversary forces as they move through open space to get to friendly cities. If U.S. air reconnaissance and strike assets are in place for locating, engaging, and attacking such adversary forces, those forces need not pass through much open terrain to suffer debilitating attrition from air attack. Aerospace power might prevent urban combat by defeating an adversary before he reached the cities that were his objective.

Developments that could hinder the use of aerospace power in preclusion strategies might include the unintended privation of innocents, surprise assaults on cities or indifference to losses caused by U.S. air attack, asymmetric adversary countermoves, and a battle for time in which the adversary enjoyed the edge in patience and stamina.

In crafting preclusion strategies aimed at circumventing such untoward developments, there is ample room for more-creative thought about involvement of joint forces than what the respective roles of air and land forces will be. In this respect, a USAF planner concerned with urban combat has recently remarked that "we can bring our technical superiority to bear only through joint operations. . . . Ground forces need the perspective of a city that is obtainable only from air and space if they are to operate there effectively. Likewise, aerospace forces may need the perspective of a city that is obtainable only from troops on the ground if they are to bring their power to bear most effectively."[18]

In all events, air-delivered weapons, if well directed, lethal, and timely, can permit ground commanders to do things differently, as well as to do different things. For example, aerospace power can fix adversary forces so that friendly ground forces can do the killing with minimal losses. It can disrupt or delay adversary strategies if friendly ground forces have not yet arrived. As a potential result, the need for an early commitment of friendly ground forces is reduced, friendly soldiers' lives are saved by substituting precision standoff attacks for ground forces in contact with the adversary, and civilian lives are saved through the use of precision to minimize collateral damage.[19]

Just as future adversaries may seek asymmetric options to negate those combat advantages the United States now enjoys in less-obstructed settings, current U.S. air and space capabilities endow decisionmakers with the option of pursuing asymmetric counterstrategies aimed at nullifying would-be enemy urban options. As a respected Australian aerospace power scholar has recently noted in this respect, "asymmetric domination enables [superior air and space assets] to strike when, where, and how they choose; and, perhaps even more importantly, to watch, listen, and know what potential enemies might do. . . . 'Asymmetric strategy' is a two-way street."[20] Apart from nuclear weapons, virtually the entire panoply of U.S. air and space potential can contribute to such preclusion operations, including not only attack platforms but also reconnaissance, surveillance, lift, and combat search-and-rescue assets. Navy carriers would have an important role to play as well in littoral contingencies. If deployed on-station in the right places when a need for prompt intervention arose, they could represent the first U.S. assets on the scene to enforce a preclusion strategy.

---

[18]Col James R. Callard, USAF, "Aerospace Power Essential in Urban Warfare," *Aviation Week & Space Technology*, September 6, 1999, p. 110.

[19]Very close coordination among force elements will be necessary to separate aircraft and weapon flight paths and to avoid fratricide. Urban operations will require a level of jointness and integration of C3I well beyond that achieved to date.

[20]Alan Stephens, *Kosovo, Or the Future of War*, RAAF Fairbairn, Australia: Royal Australian Air Force, Air Power Studies Center, Paper Number 77, August 1999, p. 19.

To be effective, any attempt to use aerospace power for avoiding an urban-combat contingency must be tied to a clear going-in strategy. U.S. decisionmakers will ask before all else, "What American interests require the insertion of U.S. ground forces into an urban jungle in the first place?" Engaging in urban combat will almost always be a matter of choice for U.S. leaders. Just because the incidence of urban challenges may be on the rise in the post–Cold War era does not automatically imply an increasing need for U.S. forces to get ensnared in them. This brings us to a review of those urban situations in which preclusionary strategies may have worked.

WHERE PRECLUSIONARY STRATEGIES MAY HAVE WORKED

As a baseline for contemplating the possibilities and associated requirements of alternative prevention strategies, it might be useful to revisit some past urban operations that did *not* occur or were substantially minimized by the fortuitous application of aerospace power.

Al Khafji

The battle of Al Khafji during the third week of Operation Desert Storm is a particularly revealing example. Once it had become clear that the allied air campaign might be open-ended, Saddam Hussein sought to draw U.S. ground forces into a slugfest that would send body bags home and turn American opinion against the war.

Toward that end, on January 29, 1991, Saddam launched a multi-pronged attack from Kuwait into Saudi Arabia in the vicinity of Border Observation Post 4 in the west and near the abandoned and unprotected coastal town of Al Khafji in the east. Whether Saddam sought to capture, reinforce, and defend Khafji or drive farther south (perhaps toward the forward logistics base at Kibrit) may never be known. Thus, we cannot know for sure that a significant urban battle was avoided. But the effectiveness of air strikes against follow-on Iraqi forces illustrates the potential for interdiction as a preclusionary strategy.

A battalion-sized element of Iraqi troops occupied the evacuated and undefended town on January 29. In the process, they unknowingly trapped two U.S. Marine reconnaissance teams. Coalition ground forces evicted the Iraqis on January 31 with minimal casualties.[21] Although aerospace power was unable to stop this initial "Battle of Khafji," it was much more effective against Iraqi follow-on forces.

Later, JSTARS aircraft detected a second wave of Iraqi columns forming up to reinforce those that had initially attacked Al Khafji. At first, the Tactical Air Control Center (TACC) did not react to these indicators of Iraqi forces on the move; its airborne sensors had been focused on areas to the west, in support of the counter-Scud operation and because Central Command's (CENTCOM's) leaders, despite the initial Iraqi foray into Al Khafji, were not expecting enemy forces to launch a major move against Saudi Arabia. However, once it became clear that a sizable ground advance was forming up on the night of January 30, the senior Central Air Force (CENTAF) officer in the TACC swung JSTARS to the east and began diverting coalition fighters to engage moving ground targets in Kuwait. Upon being apprised of the Iraqi troop activity, the JFACC saw an opportunity shaping up to engage the Iraqi column before it made contact with allied ground forces. Affirming the decision to divert coalition aerospace power from its original Air Tasking Order (ATO), he committed more than 140 aircraft against the advancing column, which consisted of two brigades from the Iraqi 3rd Armored and 5th Mechanized Divisions.[22]

The ensuing air attacks continued well into the next day before the battle was over. As a result of the timely diversion of coalition fighters, the follow-on Iraqi forces were able neither to reinforce the battalion in Khafji nor to attack south of the city. Once the dust had set-

---

[21]This occasioned one of the early misfortunes of the war, when 11 Marines were killed in two light armored vehicles that were taken out by inadvertent friendly fire, one from a Maverick missile fired by an A-10 and the other to surface fire, after the Marines had called in air strikes on the Iraqi probes.

[22]See William Murray with Wayne W. Thompson, *Air War in the Persian Gulf,* Baltimore, Md.: The Nautical and Aviation Publishing Company of America, 1995, pp. 251–253; and Rick Atkinson, *Crusade: The Untold Story of the Persian Gulf War,* Boston: Houghton Mifflin, 1993, pp. 198–213.

tled, coalition aerospace power had completely debilitated the advancing Iraqi column, forcing the survivors to beat a ragged retreat.

The hero of this story was the E-8 JSTARS, which was still in the early stages of development at the time the war began.[23] Two of these aircraft had been brought into the theater only days before the start of the war. JSTARS proved indispensable in providing the JFACC with real-time intelligence and targeting information on advancing and retreating Iraqi ground forces. Those combat aircraft the JSTARS controlled typically experienced a 90-percent success rate in finding and engaging assigned targets on the first pass. And the attacking aircraft operated so efficiently as a result that they consistently ran out of munitions before they ran out of fuel. In one instance, 80 percent of an advancing Iraqi unit was disabled before it could move into position to attack allied ground forces.[24]

The ability of the aircraft to detect and fix vehicular traffic under way with its Moving Target Indicator, while scanning large areas with its synthetic aperture radar (SAR), produced a unique synergy. In this manner, JSTARS redefined the meaning of using real-time battlespace awareness to make the most of a casebook target-rich environment. Its ability to cue allied aircraft against enemy vehicles with such deadly accuracy must have been read by the Iraqi leadership as a grim omen, since the abortive attack on the town of Al Khafji was Iraq's one and only attempt at offensive ground operations.

From then on, Iraq's forces hunkered down to absorb the coalition's punishment. The USAF chief of staff at the time, General Merrill A. McPeak, later summed up the role played by coalition aerospace power in preventing a potentially worse situation when he declared that "history may judge that the most significant battle of Desert Storm was, in fact, the one that did not happen at Al Khafji."[25]

---

[23]For an informed treatment of the E-8's concept of operations and its pivotal role in helping to preclude an urban showdown at Al Khafji, see Price Bingham, *The Battle of Al Khafji and the Future of Surveillance Precision Strike*, Arlington, Va.: Aerospace Educational Foundation, 1997.

[24]David A. Fulghum, "Desert Storm Success Renews USAF Interest in Specialty Weapons," *Aviation Week & Space Technology*, May 13, 1991, p. 85.

[25]Gen Merrill A. McPeak, USAF, *Presentation to the Commission on Roles and Missions of the Armed Forces*, Washington, D.C.: Headquarters United States Air Force, September 14, 1994, p. 101.

Kuwait City

The battle of Al Khafji was not the only recent instance in which friendly aerospace power successfully precluded or limited urban combat. The retaking of Kuwait City by coalition ground forces during Operation Desert Storm is another. Only *after* the allied air campaign had wreaked havoc on Iraqi armor, command and control, lines of communications, and, most important, the will to fight, did coalition forces launch the ground offensive that freed Kuwait. Kuwait City proper was freed with virtually no urban fighting, Iraqi forces having fled the city previously.

Southern Lebanon

Finally, we cite Israel's recurrent exposure to urban challenges since 1982 in dealing with Palestinian guerrilla operations conducted against Israel from southern Lebanon as an example of an urban experience gone bad that led to a decision to resort to aerospace power the second time around.

Israel launched Operation Peace for Galilee against Lebanon in June 1982, on the heels of a Palestine Liberation Organization (PLO) assassination attempt against Israel's ambassador to London that left him gravely wounded. That invasion by an Israeli Army division eventually reached as far as Beirut. The Israelis occupied Beirut and engaged in urban fighting for more than two months before PLO forces finally agreed to vacate the city. More than 18,000 people, mostly Lebanese civilians, were reported killed and 30,000 injured during the invasion and ensuing occupation. According to a reputable account, the invasion "drew Israel into a wasteful adventure that drained much of its inner strength and cost the IDF the lives of over 500 of its finest men in a vain effort to fulfill a role it was never meant to play."[26]

When, more than a decade later, a different Israeli government faced a comparable provocation from Palestinian terrorists operating out of southern Lebanon, the Israeli Defense Force (IDF) again struck

---

[26]Ze'ev Schiff and Ehud Ya'ari, *Israel's Lebanon War*, New York: Simon and Schuster, 1984, p. 301.

back against targets in Beirut in Operation Grapes of Wrath, which commenced on April 11, 1996, and continued for 17 days. This time, however, no IDF ground forces were committed. Instead, the IDF's concept of operations entailed a naval blockade of the ports of Beirut, Sidon, and Tyre and a steady barrage of air, artillery, and naval gunfire attacks against urban targets in Beirut and those of its environs suspected of housing Hezbollah forces. That operation, in which Israel claimed 50 Hezbollah members killed, ended on April 26 after an understanding between the two warring sides was negotiated. In all, the operation saw some 2,350 Israeli Air Force (IAF) combat sorties flown over Lebanon and not a single IDF combat fatality.[27]

WHERE AEROSPACE POWER FAILED
TO PREVENT URBAN COMBAT

In some cases, aerospace power has neither eased nor prevented urban fighting. Russia's abortive war in 1994–1996 against the breakaway republic of Chechnya offers an example.[28] Although that war mainly entailed a failed attempt by Russian ground forces to suppress a local rebellion, aerospace power played a part in providing recurrent, if ineffective, support to Russia's ground contingent, which was made up of some 40,000 troops. That contingent advanced toward the Chechen capital of Grozny from three directions on December 11, 1994, after troops and materiel were airlifted into Mozdok, just northwest of the secessionist republic, and a 2-day counterair operation was conducted to neutralize Chechnya's limited air force.

Russia's concept of operations sought to blockade the city. However, as the invasion unfolded, weather complications confronted Russian aircrews with ground fog, blowing snow, severe icing, and a heavy cloud buildup with a low ceiling and tops above 15,000 feet, making both high- and low-angle manual bombing impossible and also pre-

---

[27]Amnesty International, *Israel/Lebanon: Unlawful Killings During Operation Grapes of Wrath,* London, England, July 1996.

[28]For more on Russian air operations against Chechnya, see Benjamin S. Lambeth, *Russia's Air Power in Crisis*, Washington, D.C.: Smithsonian Institution Press, 1999, pp. 117–145.

cluding the use of electro-optical or laser-guided weapons. Instead, the Russian Air Force was forced to resort to day and night level bomb releases[29] from medium altitude (15,000–20,000 feet) against radar offset aimpoints and to inertial bombing of geographic coordinates through heavy cloud cover. The inaccuracy of those weapon deliveries resulted in many Russian losses to friendly fire.

The battle for Grozny soon degenerated into close-quarters urban combat. Yet, instead of trying to ferret out rebel defenders from their positions, Russian troops just blew up their suspected hiding places with tanks and artillery and accepted the collateral damage. Russian forces finally captured Grozny on March 7, after nearly three months of heavy fighting. For two more months thereafter, Russian aircraft continued to conduct daily attacks against the outlying road net and associated villages harboring enemy units. By that time, rebel forces had taken refuge in the surrounding mountains, from which they prevailed over Russia the following year.

In all, the war claimed the lives of four Russian Air Force pilots through the downing of two Su-25s and an Su-24. Ten Russian Army helicopters were also shot down. Most of the helicopter crews were rescued, but there were reports that at least one of the crewmembers was captured and executed by Chechen soldiers. After the war ended, a veteran's organization estimated that 4,379 Russian soldiers had been killed in the war and 703 more were missing. The Russian Air Force commander later reported that his aircraft had flown a total of 14,000 sorties since the onset of hostilities, all to no avail.[30]

Viewed in hindsight, Russia's experience in Chechnya reflected errors at all levels. To begin with, poor Russian security gave the rebels ample warning of the impending invasion, diminishing any advantage of surprise. A bigger Russian mistake was to drive into the center of Grozny with armored vehicles exposed to rebels hidden inside and atop buildings. Having failed first to encircle the city, clear an ingress

---

[29]Level bomb releases are made while flying at a constant altitude, as opposed to dive-bombing. They are appropriate if weapons are laser-guided. However, the Russians have few laser-guided weapons and the bad weather forced inaccurate radar-bombing instead.

[30]General Pyotr S. Deinekin, "Where Are We Directing the Flight of Our Birds? On the Air Force's Status and Development Prospects," *Armeiskii sbornik*, August 1996, pp. 9–12.

route, and secure a safe-escape option, the Russian military sent in some 250 unprotected tanks and armored personnel carriers. The invading tanks became separated from supporting infantry almost immediately, which made them easy prey for Chechen irregulars armed with antitank guided missiles (ATGMs) and rocket-propelled grenades (RPGs). For their part, the rebels had the advantage of mobility and of knowing the city in every detail that mattered. They used alleys and sewers to get around while the Russians remained trapped in their armored vehicles. There is little that aerospace power could have done to compensate for such flat-footed miscalculations.

However, Russian aerospace power did have at least one perceived success. Months after the heaviest fighting had ended, aerospace power was apparently the instrument of choice that finally killed the Chechen leader, Dzhokar Dudayev. In an operation that probably reflected a blend of proficiency and good luck, Russian intelligence was said to have zeroed in on Dudayev's position and transmitted coordinates to Russian ground-attack aircraft, which then fired radio-frequency homing missiles that targeted Dudayev while he was talking on a satellite field telephone. According to Russian press accounts afterwards, two missiles were electronically guided by signals bouncing between the portable phone's antenna and a relay satellite.[31]

Overall, Russian aerospace power found itself pitted against needlessly high odds as a result of poor planning. Air attacks commenced before the full ground invasion and had no clear targets, their sole intent apparently being to frighten and kill civilians. Further complicating Russia's use of aerospace power was the prohibitive weather known to afflict the northern Caucasus from November through January. The situation deteriorated at precisely the moment the initial ground assault began. After the invasion was under way, the Russian defense minister made slow and indecisive use of his air assets. Moreover, aerospace power was used not to ease the job of Russia's ground forces but, rather, to intimidate Chechen civilians and pulverize the city.

---

[31]Richard Boudreaux, "Chechens Drop Russia Talks After Leader's Death," *Los Angeles Times*, April 25, 1996.

Coordinating attack helicopter operations with those of other Russian force elements also proved to be a problem, because there was no single joint-force commander. And, since rotary-wing aviation, previously assigned to the Soviet Air Force, had been transferred to army ownership in 1990, many ground commanders had only the vaguest idea of the capabilities of helicopters and of the restrictions on their use with respect to weather, aircraft and weapons limitations, weapons range, airspeeds, and aircraft load-carrying capacity. As a result, the helicopters showed the effects of a less-than-seamless integration with the ground forces. For their part, the Chechen rebels made effective use of ambush tactics, concealing their presence and shooting from multiple directions once a helicopter entered their zone of fire. Ground-forces aviation personnel were not ready to risk taking hits from rooftop snipers. Indeed, their commander stated as the formal doctrine of his community that "urban combat is not suited to helicopters."[32]

Lessons Learned

Yet the ugliness of the battle for Grozny and Russia's ignominious defeat by the Chechen rebels should not be read to suggest that Chechnya somehow represents the wave of the future for global conflict. To begin with, the war was wholly a matter of choice for President Yeltsin. It could have been avoided altogether in favor of other approaches toward dealing with the Chechen resistance. Even taking the war as a fait accompli, Russia could have used its air assets more intelligently.

Because of chronic underfunding and a consequent lack of aircrew training and proficiency, Russia's Air Force was anything but ready to go to war. It also lacked the fielded technology in the form of a JSTARS-equivalent that might have been used to enforce a no-drive zone in and around Grozny. Nevertheless, from a weather perspective, Russian air commanders could have insisted on a better time to commence air operations. They also could have forgone carpet bombing, which only ensured the everlasting enmity of the surviving civilian populace. Instead of being used as flying artillery, aircraft

---

[32]Paul Beaver, "Army Aviation in Chechnya," *Jane's Defense Weekly*, June 10, 1995, p. 79.

could have isolated the central part of the city from outlying sectors, provided better force protection for Russian troops deployed within the city limits, and shown greater concern for avoiding collateral damage. For their part, the Russian Army's attack helicopters could have avoided flying repeatedly into enemy kill sacks in urban canyons, and Chechen radio and television stations could have been jammed rather than bombed into silence.

In these and other ways, Russia's aerospace power could have been used in lieu of an immediate ground invasion, which could have been deferred until the weather and enemy resistance were more manageable.

The Second Chechen Battle

Three years later, the Yeltsin government had another opportunity to "get it right" when it went to war a second time against Chechnya in August 1999, this time in response to a spate of apartment bombings in the Moscow region that it alleged to be the work of Chechen terrorists. In that second attempt to subdue the Chechen rebellion and impose Russian control over the separatist republic, both the Yeltsin leadership and Russia's military commanders showed evidence of learning from past experience when they combined massed firepower and a refusal to get drawn into urban combat into what amounted to a siege strategy centering on annihilating the Chechen resistance from a safe distance. This time, Russian commanders avoided close-in urban engagements and opted instead to use their air and artillery assets in standoff attacks against assumed rebel targets in the city. In late December, Russian ground forces did enter the city, replaying the earlier experience of bloody fighting and large loss of life. Fortunately for the Russians, ground fighting in the city was fairly short-lived this time around. This 6-month siege eventually led to a victory for Moscow in late January 2000, after Russian firepower had finally reduced Grozny to a smoldering shell and driven what remained of the rebel forces to flee the city and continue their resistance from the surrounding mountains.

This time, as if to emulate Israel's resorting to standoff operations during the second challenge it faced from southern Lebanon, the Russians avoided committing large numbers of their infantrymen to early urban combat and sought instead to surround Grozny, sever the supply routes to the surrounding mountains emanating from Grozny, and block the southern passes toward Georgia through which fleeing rebels might seek to escape. Those in command admittedly chose a better time of year, from a weather perspective, to commence air operations, and they definitely sought to employ aerospace power as a substitute for a ground invasion. Moreover, although the Russians had suffered 1,500 or more combat fatalities by the time the heaviest fighting had ended, they succeeded in minimizing their losses to friendly fire this time by taking determined care not to insert their own troops into areas in and around Grozny where Russian bombs would be falling.

However, this second Russian experience in Chechnya was by no means any more an effective use of aerospace power in mitigating urban combat than the first. On the contrary, rather than using aerospace power to minimize collateral damage, the Russians applied it both indiscriminately and even wantonly to destroy civilian structures outright and to terrorize the civilian population. In effect, they conducted combat operations not *in* an urban environment but rather *against* one. Through some 5,000 Su-24 and Su-25 sorties all together, they used their aerospace power once again merely as flying artillery, and they showed no significant improvements in tactics or proficiency over their earlier performance in 1994–1995.[33] In the process, they lost to enemy fire at least two Su-25 attack aircraft, an Su-24MR reconnaissance jet, and two helicopters—one an Mi-24 as it was performing a search-and-rescue mission to extract the crewmembers of a previously downed Mi-8.[34]

WHEN CIRCUMSTANCES LEAVE NO CHOICE

As the preceding discussion has sought to show, aerospace power should be viewed, first of all, *not* as a means for prosecuting urban combat more effectively but, rather, as an instrument for preventing

---

[33]David Fulghum, "Air War in Chechnya Reveals Mix of Tactics," *Aviation Week & Space Technology*, February 14, 2000, p. 77.

[34]Alexei Komarov, "Chechen Conflict Drives Call for Air Force Modernization," *Aviation Week & Space Technology*, February 14, 2000, p. 80. Another 24 Russian aircraft reportedly sustained battle damage but managed to return safely to their nearby bases just across the Chechen border.

or controlling urban strife before it degenerates into the kinds of situations typified by Somalia and Chechnya. As airmen ponder the various ways in which U.S. air and space assets can help a CINC handle the worst kinds of urban contingencies, they should educate CINCs in the many ways in which those same assets, used in a timely manner, can make handling such contingencies unnecessary.

At the same time, airmen must take care to avoid the opposite extreme: defining away the urban-combat challenge by fiat by assuming that aerospace power's comparative advantages for prevention will *always* be applicable to the needs of a CINC. Adversaries will be tempted to pursue urban options as a means of end-running U.S. technology advantages; therefore, urban contingencies can arise almost anywhere along the conflict spectrum, from humanitarian airlift and assistance through humanitarian intervention, NEOs, and small-scale shooting contingencies, to major theater wars. In such contingencies, adversaries will seek to gain the safety of hardened positions, use civilian populations and facilities as shields, and generally make the most of urban settings for concealment and deception. They also will try to prevent the use of airlift terminals and lines of communication, maximize U.S. and allied casualties, and render conflicts drawn-out enough to stretch U.S. endurance to the breaking point.

Accordingly, airmen must recognize and accept that aerospace power is not a universally applicable tool for solving every urban-military-operation challenge that might arise. Insofar as urban fighting typically stems from internal conflicts in which all of the players are in place before U.S. forces have arrived on the scene, preemptive strategies that might work well enough on a pristine playing field will be ruled out before the fact in most urban challenges confronting U.S. planners. There will be times when exigencies of the moment—humanitarian, political, or military—will leave U.S. leaders with no choice but to engage in urban operations: for example, finding and seizing weapons of mass destruction (WMD) facilities in defended urban settings, rescuing friendly hostages or prisoners of war (POWs), extracting noncombatants, providing humanitarian assistance under fire, and supporting friendly governments against urban insurgents. Under such contingent circumstances, U.S. aerospace power—as for all U.S. force elements—will have little choice but to react as best it can.

As a Tool for Joint Force Commanders

That said, although U.S. air and space assets may at present have only a limited ability to perform certain tasks in the messier sorts of future urban contingencies, they may eventually offer the promise of greater versatility. For example, current and projected aids to battlespace awareness afforded by U.S. air and space assets give U.S. joint force commanders tremendous advantages over potential opponents in their ability to integrate and orchestrate various force elements to seek the greatest efficiencies in a given operational situation.

Circumstances can be imagined in which the possibility of high-intensity urban combat would be a part of the operational situation from the outset. For example, if war again breaks out on the Korean Peninsula, friendly and enemy forces will be in contact from the opening moments of fighting. Granted, North Korean troops might bypass Seoul rather than attempt to seize and occupy it. On the one hand, their leaders could endeavor to bottle up the city and isolate it rather than hazard the perils of going downtown with armor and mechanized infantry. On the other hand, the possibility of South Korean towns being overrun by the enemy, or of portions being seized by enemy agents and previously inserted special operations forces, could create a need for allied forces to starve out or root out the enemy from cities at the outset of a counterattack.

Moreover, should urban operations become the norm, aerospace power need not *always* be defensive and supporting. It can also be offensive, featuring different problems and opportunities. For example, in the context of a conflict involving significant casualties, rules of engagement may be much more permissive and allow a CINC broad discretion in engaging adversary forces in urban areas. The threat of urban engagements would not necessarily have to slow down a counterattack's rate of advance, and the CINC could call on offensive aerospace power as the option of choice for performing the needed destruction.[35]

---

[35]Conversation with GEN John H. Tilelli, Jr., USA, Commander in Chief of U.S. Forces in Korea (CINCUSFK), Seoul, Korea, June 9, 1999.

The Limits of Aerospace Power in Urban Operations

What aerospace power cannot currently do very well is impose sustained force against urban insurgents or discriminate between adversaries and innocents in an urban setting. Accordingly, aerospace power will tend to perform best when the desired outcome involves affecting adversary behavior rather than seizing and holding urban terrain. The challenge will be to minimize the risk to friendly combatants, employing ground forces as a last resort and only after aerospace power has adequately paved the way for them.

SUMMARY

In sum, when aerospace power cannot preclude urban operations, it can seek to mitigate their worst aspects. When it cannot mitigate, it can strive to control by isolating adversary combatants, then supporting the CINC's efforts to engage them in detail.[36] As difficult as those tasks may be compared with the cleaner preclusion options considered above, both existing capabilities and emerging technologies and concepts of operations offer a significant prospect for U.S. aerospace power to help defeat unconventional opponents who would seek to turn urban situations to their advantage. It is to that emerging challenge that the remainder of this report is addressed.

Before turning to new concepts to better exploit aerospace power where an urban operation cannot be avoided, we first look at a variety of constraints on air operations in urban settings, beginning in Chapter Three with a look at the urban legal and political environment.

---

[36]This line of reasoning is further developed in Colonel Kevin Kennedy, USAF, *MOUT: An Airman's Perspective*, briefing to a conference on "The Role of Aerospace Power in Joint Urban Operations" sponsored by Air Combat Command, Air Force Special Operations Command, and Headquarters USAF (AF/XPX), Hurlburt Field, Fla., March 23–24, 1999.

Chapter Three

LEGAL AND POLITICAL CONSTRAINTS ON URBAN AEROSPACE OPERATIONS

INTRODUCTION

Recent U.S. and coalition operations in the Balkans and elsewhere have been marked by heated controversy over target selection. Demonstrating the difficulty of balancing the often-competing concerns of avoiding collateral damage, minimizing risk of U.S. casualties, and maximizing military effectiveness, they show that many of the most limiting constraints on future U.S. urban military actions are going to be legal and political, not technological or operational.

To be sure, legal and political constraints are not independent of technological and operational constraints. Expanded capabilities or new operational concepts may provide the means of reducing or avoiding collateral damage and enhancing force protection, and they may inform the public perceptions that drive legal and political constraints. However, in planning for urban operations, the most salient limitations on U.S. military action are often those the military places on itself by adhering to international legal norms and restrictive rules of engagement (ROE) to satisfy public and diplomatic pressures.

Urban environments may pose enormous difficulties for those planning and conducting military operations within the boundaries of international law and self-imposed constraints on the use of force. The speed and agility of aerospace power, combined with its ability to deliver firepower precisely and with fairly low risk to U.S. personnel across the spectrum of conflict, often make it the military instrument of choice for decisionmakers. However, the heightened risk of collateral damage when operating in urban environments partially offsets U.S. technological superiority and provides adversaries with expanded opportunities to exploit U.S. adherence to certain norms. As a result, the urban-combat options available to planners and decisionmakers are generally far narrower than the domain of the feasible.

Effective planning for future urban operations, with its consideration of new technologies and operational concepts, requires first an appreciation of the factors that will govern whether and how those technologies and concepts will be used. This chapter aims to place urban aerospace operations in their legal and political context, thereby laying the foundations for assessing USAF urban-combat capabilities across the range of potential tasks. It gives particular attention to urban MOOTW, because air operations in this context are especially difficult (owing to the inherent constraints associated with MOOTW) and may be common in the near future. But the issues presented in this chapter apply to conventional combat operations as well. The chapter also draws on air campaign planning experiences during the Vietnam and Persian Gulf Wars and, more recently, NATO operations over Kosovo—operations that, while lying outside the focus of this volume, spotlight the most pertinent competing pressures that constrain planning of all air operations.

LEGAL CONSTRAINTS ON URBAN
AEROSPACE OPERATIONS

The body of norms regulating the conduct of states and combatants engaged in armed hostilities is the law of armed conflict.[1] International law generally derives from both treaties (conventions and agreements among states) and custom (behavioral norms that become widely recognized among states as binding). Contemporary law of armed conflict draws heavily from the Hague Conventions,

---

[1]The term "law of war" is often used interchangeably with "law of armed conflict," even though the legal requirements placed on parties sometimes depend on the type of conflict or operation being waged. This chapter is concerned with the legal norms that apply across the spectrum of conflict and, for clarity's sake, employs throughout the term "law of armed conflict."

negotiated at the peace conferences of 1899 and 1907, and the Geneva Conventions, as well as from numerous agreements that limit the means and conduct of hostilities.[2] Together with norms that have evolved out of customary practice among states,[3] these various sources of law aim to reduce damage and the suffering of combatants and noncombatants during conflict.

The United States has a strong interest in upholding international norms, which tend to be stabilizing forces and increase the predictability of state actions. Beyond a general commitment to upholding the law of armed conflict, U.S. planners place a premium on the perceived legitimacy of military operations. That legitimacy often turns, in part, on the perceived legality of those operations, although, as discussed below, demands to maximize U.S. force protection and military effectiveness sometimes create strong countervailing pressures that must be balanced. Planners strive to maintain support among three sets of audiences—the domestic public, the international community, and some parties local to the areas of operations; adherence to international law can bolster this support.

While interpretations of some international legal provisions shift or remain unsettled, many of the basic principles (e.g., proportionality) embodied in the law of armed conflict have, over time, been internalized in the highest levels of strategic planning down to the lowest levels of tactical decisionmaking by individuals, when commitment to legal norms helps motivate and sustain the morale of U.S. servicemen.[4] Prior to and during operations, legal advisers and military Judge Advocate General (JAG) officers have played a variety of roles

---

[2]For example, the 1925 Geneva Protocol for the Prohibition of the Use in War of Asphyxiating, Poisonous or Other Gases and of Bacteriological Methods of Warfare prohibits the use of some types of chemical weapons.

[3]The 1977 Protocols Additional to the Geneva Conventions of 1949 (often referred to simply as Protocol I and Protocol II) spell out specific sets of rules to govern international and internal conflicts. The United States has not ratified the Protocols; it has declared its intention to be bound by them to the extent that they reflect customary law. Michael J. Matheson, "The United States Position on the Relation of Customary International Law to the 1977 Protocols Additional to the 1949 Geneva Conventions," *American University Journal of International Law and Policy*, Vol. 2, 1987, pp. 419–431.

[4]John G. Humphries, "Operations Law and the Rules of Engagement," *Airpower Journal*, Vol. 6, No. 3, Fall 1992, pp. 38–39, describes how international legal norms have been internalized by military planners and operators, particularly since the Vietnam War.

in ensuring compliance with international law—roles that have gained prominence in the past decade.[5]

The body of international law regulating armed conflict is intricate, and many of its key provisions remain contested among states, international organizations, and scholars. But much of the law of war can be reduced to several key concepts around which there is near-consensus: military necessity, humanity, proportionality, and distinction (or discrimination).

International regulation of armed conflict begins with the principle of *military necessity*, "the principle which justifies measures of regulated force not forbidden by international law which are indispensable for securing the prompt submission of the enemy, with the least possible expenditures of economic and human resources."[6] In pursuing military victory, however, parties are also governed by the principle of *humanity*, which forbids the infliction of injury or destruction not necessary to the achievement of legitimate military purposes.

To a degree, the principles of military necessity and humanity complement each other, both reflecting the notion of economy of force. Yet, there is also a tension between them that the law of armed conflict seeks to mediate, between allowing sufficient military flexibility to subdue an enemy and restricting that flexibility to reduce the destructive impact of conflict. Many of the specific rules contained in the law of armed conflict attempt to balance, on the one hand, the latitude necessary for military forces to carry out their functions, with, on the other hand, a desire to minimize human suffering.[7]

---

[5]More and more, international law experts and military judge advocates are paying particular attention to review of target lists and promulgation of rules of engagement (ROE).

[6]U.S. Department of the Air Force, *International Law—The Conduct of Armed Conflict and Air Operations*, Air Force Pamphlet 110-31, November 1976, para. 1-3(a)(1).

[7]The precise formulation of these key principles varies. U.S. Department of the Navy, *The Commander's Handbook on the Law of Naval Operations*, NWP1-14M/FMFM 1-10/COMDTPUB P5800.7, October 1995, para 8.1, for example, enumerates the following three fundamental principles of the law of armed conflict that regulate targeting:

- The right of belligerents to adopt means of injuring the enemy is not unlimited.
- It is prohibited to launch attacks against the civilian population as such.

Together, the principles of military necessity and humanity underlie the rule of *proportionality*, which demands that parties refrain from attacks, even against legitimate military targets, likely to cause civilian suffering and damage disproportionate to the expected military gain.[8] A classic example of this rule holds that if a force advancing through a town encounters a single enemy sniper firing from atop a hospital, the force is prohibited from demolishing the entire building, especially if less extreme options are available, because the civilian harm would far outweigh the military advantage attained.

Embedded in these principles and rules is the idea of *distinction* (or *discrimination*) between military and civilian persons or property. Planners and commanders are obligated to distinguish between military and civilian targets, restricting their attacks to the former only. In broad terms, international law prohibits attacks on civilian populations as such, as well as acts or threats of violence having the primary purpose of spreading terror among the civilian population.[9] Furthermore, operations are to be directed exclusively at *military objectives*, defined as "those objects which by their own nature, location, purpose, or use make an effective contribution to military action and whose total or partial destruction, capture, or neutralization in the circumstances ruling at the time offers a definite military advantage."[10]

Air Force Pamphlet 110-31, *International Law—The Conduct of Armed Conflict and Air Operations*, instructs that, in applying international legal limits to air attacks, planners must take the following precautions:

- Distinctions must be made between combatants and noncombatants, to the effect that noncombatants be spared as much as possible.

[8]Article 51(5)(b) of Protocol I prohibits "an attack which may be expected to cause incidental loss of civilian life, injury to civilians, damage to civilian objects, or a combination thereof, which would be excessive in relation to the concrete and direct military advantage anticipated." The United States has accepted this provision as reflecting international law. Matheson, 1987, p. 426. Although this principle is almost universally regarded internationally as law, its precise meaning remains elusive, in part because of the inherent difficulties in measuring, and then weighing, expected military gain and civilian harm.

[9]AFP 110-31, 1976, p. 5-7.

[10]AFP 110-31, 1976, p. 5-8. A virtually identical definition is contained in the 1977 Protocols Additional to the Geneva Conventions of 1949 (Article 52 of Protocol I).

(a) Do everything feasible to verify that the objectives attacked are neither civilians nor civilian objects . . .

(b) Take all feasible precautions in the choice of means and methods of attack with a view to avoiding, and in any event to minimizing, incidental loss of civilian life, injury to civilians, and damage to civilian objects; and

(c) Refrain from deciding to launch any attack which may be expected to cause incidental loss of civilian life, injury to civilians, damage to civilian objects, or a combination thereof, which would be excessive in relation to the concrete and direct military advantage anticipated.[11]

Note that these precautions embody the principles outlined above: discrimination (section a), humanity (section b), and proportionality (section c).

Aside from the general rule prohibiting direct attacks on civilian objects, narrow rules also proscribe attacks on specific objects granted special protection. The 1949 Geneva Conventions protect hospitals and other medical assets. Religious and cultural buildings and monuments are also promised special protected status under international law.[12]

These provisions are not open and shut but give way to contention and further interpretation. Authorities disagree on whether planners and operators legally must always select the weapon, from among those capable of destroying a target, that poses the *least* risk of collateral damage and civilian injury when operating in highly populated areas. The principles of humanity and distinction give rise to the consensus view prohibiting weapons that cause superfluous

---

[11]AFP 110-31, 1976, p. 5-9. These requirements mirror almost verbatim the provisions in Protocol I, Article 57. In attacking even legitimate military targets, commanders may also be obligated to issue warnings to civilians within their vicinity. This requirement is long-standing, and codified in Hague Regulations, though with an important caveat for cases of assault, when advanced warning would spoil tactical surprise. Thus, "[g]eneral warnings are more frequently given than specific warnings, lest the attacking force or the success of its mission be jeopardized."

[12]This status was codified in Article 27 of Hague Regulations Respecting the Laws and Customs of War on Land, 18 October 1907, and Article 5 of Hague Convention No. IX Concerning Bombardment by Naval Forces in Time of War, October 18, 1907.

injury (e.g., poisoned projectiles; dum-dum bullets) or are completely incapable of discrimination (e.g., WWII German V-1 and V-2 rockets).[13] Some scholars and organizations argue that, beyond these minimal threshold prohibitions, an attacker must choose the means and methods for minimizing the risk of incidental civilian damage to the greatest extent feasible.[14] Under this interpretation, for example, U.S. forces would always be legally obligated to use precision-guided munitions against urban targets. Since the Vietnam War, U.S. forces have virtually always done so for reasons related to politics or military effectiveness, but the U.S. military generally opposes this more restrictive legal interpretation.[15]

Another increasingly contentious issue involves choices between weapon systems, particularly standoff weapons, that may trade off increased force protection for a heightened risk of collateral damage. Again, the U.S. military favors a liberal interpretation of weapon-selection duties, one that permits an extremely high level of force protection so long as an appropriate level of accuracy is still ensured.[16]

In the remainder of this section, we focus on the reciprocal legal duties and the defender's obligations and the international legal challenges of urban environments.

---

[13]AFP 110-31, 1976, p. 6-2; U.S. Department of the Navy, 1995, p. 9-1.

[14]See, for example, Michael Bothe, Karl Josef Partsch, and Waldemar A. Solf, *New Rules for Victims of Armed Conflicts*, The Hague: Martinus Nijhoff Publishers, 1982, p. 364; Middle East Watch, *Needless Deaths in the Gulf War: Civilian Casualties During the Air Campaign and Violations of the Laws of War*, New York, 1991, pp. 126–127.

[15]This debate is summarized in Ariane L. DeSaussure, "The Role of the Law of Armed Conflict During the Persian Gulf War: An Overview," *Air Force Law Review*, Vol. 27, 1994, pp. 60–61.

[16]According to *The Commander's Handbook on the Law of Naval Operations*, for example, "[m]issiles and projectiles with over-the-horizon or beyond-visual-range capabilities are lawful, provided they are equipped with sensors, or are employed in conjunction with external sources of targeting data, that are sufficient to ensure effective target discrimination."

Reciprocal Legal Duties and the Defender's Obligations

So far, this outline of legal constraints has been one-sided; it has focused on regulating the *attacker's* actions.[17] Because the attacker generally has an array of options for when, where, how, and how much it employs destructive force, the law of armed conflict places on it the above-mentioned responsibilities. But the international legal regime also places corresponding duties on the *defender*.

Efforts during the past several decades to codify the law of armed conflict have emphasized the reciprocal duties of attackers and defenders. Article 58 of Protocol I demands that parties endeavor to segregate military objectives from their civilian population and take steps to protect civilians from the dangers of military operations.[18] Article 51 also provides that the "presence or movements of the civilian population or individual citizens shall not be used to render certain points or areas immune from military operations, in particular in attempts to shield military objective from attacks or to shield, favour or impede military operations."[19]

Exploiting the discrimination requirement placed on attackers by deliberately commingling civilians with military targets violates the

---

[17]The law of armed conflict, particularly as applied to air operations, often speaks in terms of "attacker" and "defender." Because this document analyzes constraints on U.S. air operations in urban environments, the former, generic term is assumed to apply to U.S. forces, while the latter describes adversaries' obligations and actions.

[18]The same principle applies to specially protected sites. For example, Article 19 of the Geneva Convention for the Amelioration of the Condition of the Wounded and Sick in Armed Forces in the Field (1949) establishes that "the responsible authorities shall ensure that the said medical establishments and units are, as far as possible, situated in such a manner that attacks against military objectives cannot imperil their safety."

[19]Protocol I, article 51(7). This admonition is similarly articulated in Air Force Pamphlet 110-31 (1976, p. 5-8), which explains:

> The requirement to distinguish between combatants and civilians, and between military objectives and civilian objects, imposes obligations on all the parties to the conflict to establish and maintain the distinctions. . . . Inherent in the principle of protecting the civilian population, and required to make that protection fully effective, is a requirement that civilians not be used to render areas immune from military operations. Civilians may not be used to shield a defensive position, to hide military objectives, or to screen an attack. . . . A party to a conflict which chooses to use its civilian population for military purposes violates its obligations to protect its own civilian population. It cannot complain when inevitable, although regrettable, civilian casualties result.

basic principles of the law of armed conflict. Note, however, that a defender's violation of these principles—for example, its deliberate placement of civilians in the vicinity of military targets or its use of specially protected sites to house weapons—does not eliminate all legal obligations on the attacker. Among other things, an attacker would generally remain bound by proportionality principles and obliged to refrain from attacks likely to result in civilian damage that is excessive in relation to military gain. Nevertheless, the relative protections normally granted those civilian persons and objects is weakened. As elaborated below, adversaries, especially those that show little regard for international law and do not face political or diplomatic pressures similar to those faced by the United States, have exploited the asymmetry of constraints for strategic or tactical gain, with some success.

The International Legal Challenges of Urban Environments

Compliance with the principles of discrimination and proportionality is confounded by the structure and organization of urban centers. From a planning viewpoint, these principles contain a foreseeability element: Planners must consider collateral damage and likely injury to noncombatants or civilian property and must take reasonable actions to avoid or minimize these effects. By connecting and tightly packing both military and civilian resources, not only may the urban environment increase the chances that military attacks will harm civilians, but it may increase the likelihood that even precise attacks can unleash substantial secondary effects on the urban population. For example, a perfectly executed attack on a power plant might not harm a single civilian directly, but thousands could be harmed indirectly through effects related to the loss of electrical power in the city.

The density of civilian populations in urban areas may increase the chances that even accurate attacks will injure noncombatants. In addition, the collocation of military and civilian assets in urban environments multiplies the chances that military attacks will cause unintended, and perhaps disproportionate, civilian damage. The proximity of civilian and military targets in urban environments exists in the horizontal dimension (military and civilian structures situated side by side) as well as the vertical dimension (military and civilian assets stacked one above the other, within the same structure).

Horizontal proximity of civilian and military sites raises both the possibility that an attack will accidentally hit nearby civilian buildings or the possibility that a direct hit on a military site will damage adjacent civilian ones. A primary objective for U.S. forces during the early phases of Operation Just Cause was the Panamanian Defense Force (PDF) general headquarters, the *Comandancia,* located in the middle of a poor Panama City neighborhood (*El Chorrillo*). During U.S. shelling of the headquarters and subsequent efforts to squelch sniper fire, several fires broke out and spread through nearby civilian residences, leading the human rights organization Americas Watch to conclude that "inadequate observance of the rule of proportionality resulted in unacceptable civilian deaths and destruction,"[20] a conclusion disputed by other post-operation analyses. Most urban air operations were conducted with direct line-of-sight precision weapon platforms, which were more accurate than indirect-fire weapons, thereby reducing the risk and extent of collateral damage.[21] Even with these precautions in place, civilian injury and damage were extensive.[22]

During Operation Desert Fox in December 1998, planners avoided bombing some facilities that contributed to Iraq's chemical weapons program because of the possibility of releasing toxins within Baghdad and the effects of those toxins.

Regardless of whether these targeting restrictions were strictly required from a legal standpoint, they illustrate that civilian and military sites need not be immediately adjacent to complicate decision-making that seeks to avoid collateral damage.

---

[20]Americas Watch, *The Laws of War and the Conduct of the Panama Invasion,* May 1990, pp. 16–21.

[21]AH-64 attack helicopters and AC-130 gunships, both with direct line-of-sight weapons and night-vision capability, were used against the *Comandancia.* John Embry Parkerson, Jr., "United States Compliance with Humanitarian Law Respecting Civilians During Operation Just Cause," *Military Law Review,* Vol. 133, 1991, p. 54.

[22]Most estimates put the total number of civilian deaths resulting from the Panama invasion between 220 and 300. (Parkerson, 1991, p. 55). Americas Watch (1990, p. 19) reported that the attack on *La Commandancia* left about 15,000 persons homeless and resulted in 50–70 civilian deaths.

The Gulf War Al Firdos bunker incident demonstrates how proximity of military and civilian targets can operate in a vertical dimension, especially in an urban environment. On the night of February 13, 1991, U.S. F-117 strikes destroyed the bunker, a building that intelligence gatherers had identified as a command and control facility. The true nature of the facility remains disputed, but on the night it was destroyed it housed families of government officials in its upper levels; the strikes thus resulted in dozens of civilian deaths. Vertical proximity creates similar potential problems for neutralizing an urban sniper without harming civilians in rooms above or below, as well as beside, his.

The legal regime recognizes the difficulty of military decisionmaking in war. It thus obligates planners and commanders to base their decisions on information available at the time, not in perfect hindsight. The fact that civilian and military targets may be stacked on top of each other in urban environments complicates the assessment of potential civilian risk in attacking certain sites, as well as the ability, even after information is gathered, to destroy only military targets. As elaborated below, both horizontal proximity and vertical proximity of military and civilian targets present adversaries with opportunities to exploit legal and political constraints to immunize legitimate targets from attack.

Particular difficulties emerge from the collocation of civilian and military assets in urban environments when air defenses are concentrated near key targets. Not only does the emplacement of air defense systems or even the possession of hand-held anti-aircraft weapons by local forces in densely populated areas compound the problem of civilian-military asset mingling, but it can increase the chance of civilian damage resulting from air attacks on military targets, because attacking aircraft may now be forced to take evasive actions, jettison ordnance, or operate at higher altitudes. As Hays Parks has explained, "The purpose of enemy defenses is not necessarily to cause aircraft losses; the defender has accomplished his mission if he makes the attacker miss his target."[23]

---

[23]W. Hays Parks, "Air War and the Law of War," *Air Force Law Review,* Vol. 32, 1990, p. 191.

Urban environments contain shared military-civilian resources and house dual-use facilities. The military and civilian population often utilize common power sources, transportation networks, and telecommunications systems. Distinguishing between military and civilian infrastructure is sometimes difficult—especially support systems that provide basic needs, such as electricity. It may be impossible to destroy or disrupt only those portions servicing the military. During crises, for example, the military is generally the priority user and can be expected to utilize any residual capacity. Attacks on shared infrastructure can also have large reverberating effects on the civilian population, giving rise to concerns about proportionality.[24] Planners sometimes view the dual-use nature of infrastructure systems opportunistically, because military usage arguably legitimizes these systems as targets, even though it may in fact be the incidental effects on the civilian population that planners hope to manipulate. As a result, the United States tends to favor liberal legal interpretations of "military objective" regarding dual-use facilities.[25]

Some of the most vocal criticism of Operation Desert Storm has surrounded air attacks on the Iraqi electrical system. Air campaign planners sought to degrade Iraq's electric-power generation and distribution capabilities during early phases of the operation to disrupt air defenses and command networks. Air planners recognized that these attacks would deny electricity to the Iraqi populace as well; to some degree, civilian deprivations were intended as part of the overall air strategy to compel the regime's capitulation.[26] Post-war accounts suggest that, to achieve immediate military objectives without subjecting the population to prolonged hardship, planners sought to avoid destroying those elements of the electric system that

---

[24]Some disagreement exists about how to calculate adverse civilian effects of attacks on military targets. One view holds that planners must consider the long-term, indirect effects of attacks on a civilian population; the U.S. military adheres to a narrower interpretation, emphasizing only direct civilian injuries or deaths. During operational planning, when target lists are reviewed for compliance with international law, much greater emphasis is typically given to immediate collateral-damage effects.

[25]"When objects are used concurrently for civilian and military purposes, they are liable to attack if there is a military advantage to be gained in their attack." U.S. Department of Defense, *Conduct of the Persian Gulf War*, Final Report to Congress, Washington, D.C.: U.S. Government Printing Office (GPO), 1992, p. 613.

[26]Barton Gellman, "Allied Air War Struck More Broadly in Iraq," *Washington Post*, June 23, 1991, p. A1.

would require long-term reconstruction. In that sense, they empha-sized discrimination in a temporal, rather than geographical, di-mension, by trying to minimize potential lingering civilian effects long after the conflict.

However, the interconnectedness of resource systems in a modern society means that attacks against certain elements can have unex-pected ripple effects. As one post-war analysis of these strikes ex-plained: "Unfortunately, it is simply not possible to segregate the electricity that powers a hospital from 'other' electricity in the same lines that powers a biological weapons facility."[27] In this case, the loss of power-generating facilities disrupted irrigation, sewage, and medical systems, contributing to massive outbreaks of waterborne diseases and other public health crises (some post-war studies recorded a civilian death toll perhaps surpassing 100,000 as a result of these effects).[28]

The dilemmas stemming from shared civilian-military resources can be expected to increase, as greater parts of the world modernize and develop networked infrastructure systems. Some military theorists welcome this trend, viewing these systems as vulnerable to U.S. aerospace power and their destruction or degradation possibly allowing planners to bypass the adversary's fielded military forces by influencing the adversary populace and its leadership decision-making.[29]

Even if operational concepts directed at disrupting these systems pass legal scrutiny, political constraints may limit their availability to planners.

---

[27]Daniel T. Kuehl, "Airpower vs. Electricity: Electric Power As a Target for Strategic Air Operations," *Journal of Strategic Studies*, Vol. 18, No. 1, March 1995, p. 254.

[28]For a critical account of coalition attacks on the Iraqi electric system and its after-effects, see Middle East Watch, 1991, pp. 171–193. Note that the long-term effects of these attacks resulted, in part, not only from international sanctions but also from re-source-allocation decisions by the Iraqi government.

[29]See, for example, John A. Warden III, "The Enemy As a System," *Air Power Journal*, Vol. IX, No. 1, Spring 1995, p. 49, who argues that "[u]nless the stakes in the war are very high, most states will make desired concessions when their power-generation system is put under sufficient pressure or actually destroyed."

POLITICAL CONSTRAINTS ON URBAN OPERATIONS

As just outlined, the law of armed conflict imposes obligations on attackers and defenders to take precautions to reduce the risk of collateral damage and civilian injury. The risk of such damage from air operations may be magnified in those urban settings where military and civilian assets are collocated and difficult to distinguish. As a result, legal constraints on air operations will often be most tightly felt by planners and operators in urban environments.

These legal obligations are supplemented by an additional set of constraints on planners that are driven by political forces. Public and coalition sensitivity to friendly casualties and collateral damage often reduces operational flexibility more severely than does adherence to the international law of armed conflict. They may also push in opposite directions. For example, a requirement that U.S. ground-attack aircraft fly at higher altitudes to minimize risks to crews could make it harder to identify vehicles or personnel on the ground and, under some circumstances, might lead to unintended attacks on civilians. As happened in Kosovo, flying that high resulted in misidentification of civilian-vehicle traffic as an enemy. Flying at lower altitudes might have allowed pilots to recognize the convoys as civilian, but at a much greater risk to the aircrews.

Political constraints derive from the need to maintain certain minimum levels of support for military operations among three audiences: the domestic public, the international community (most notably, the United States' major and regional allies), and the local population in the conflict area. The relative weight of these audiences' opinion on U.S. decisionmaking varies considerably with context and type of operation. For example, when vital U.S. interests are at stake, decisionmakers are less likely to adapt operations to placate international dissent; when peripheral interests are at stake, the relative importance of diplomatic backlash naturally rises and decisionmakers will tailor operations accordingly. During full-scale combat operations, the demands of the local populace will typically concern U.S. decisionmakers and planners less than they will during peacekeeping or humanitarian operations, in which perceived impartiality and maintaining consent of factional parties may be critical to success. Even when U.S. vital interests are at issue, these pressures affect strategic decisions about when and whether to conduct military operations at all, as well as operational decisions including choice of forces, weapons, and ROE.

In this section, we look at issues that affect the domestic public—sensitivity to American casualties—and the international community and local population in the conflict area—sensitivity to collateral damage and civilian suffering—as well as operational decisions.

Sensitivity to American Casualties

Today, U.S. military operations are planned and conducted with high sensitivity to potential U.S. casualties. Policymakers and planners generally fear that U.S. casualties will—or at least may—erode support for sustained operations. Force protection is therefore paramount in designing operations.

Contrary to the predictions of those who saw Desert Storm as putting the Vietnam experience behind, the low American death total in the Gulf War relative to Vietnam likely raised public expectations of "bloodless" foreign policy, as well as the perception among policymakers and military planners that public expectations in this regard have risen. The further erosion of already-fragile American public support that followed the October 1993 deaths of 18 U.S. servicemen in Mogadishu suggests the strong pull that casualties can exert on U.S. policy. The extended deployment of U.S. ground forces to enforce the Dayton peace accords in the former Yugoslavia only confirms this tendency: Unlike troops of other NATO partners, U.S. troops patrol in convoys and have avoided actions likely to provoke hostile responses from local factions.[30]

Although a number of empirical studies have shown that the effects of U.S. casualties on public support depend heavily on a number of other contextual variables and factors—for example, support is likely to erode with casualties when public views victory as unlikely[31]—this

---

[30]Edith M. Lederer, "Tuzla Off Limits to Off-Duty U.S. Troops," *Detroit News*, February 20, 1997, p. A12.

[31]For such conclusions and evidence drawn from other studies, see Eric V. Larson, *Casualties and Consensus: The Historical Role of Casualties in Domestic Support for U.S. Military Operations*, Santa Monica, Calif.: RAND, MR-726-RC, 1996. Larson's study showed that, in a number of past cases, support for a military operation declined

sensitivity affects policy and planning decisions both before and during operations, when concern for potentially adverse public reactions weighs heavily.

Adversaries often view casualty sensitivity as a critical element of the United States' "center of gravity": its political will to sustain operations. Somali warlord Mohammed Farah Aideed reportedly told Robert Oakley, the U.S. special envoy to Somalia during the U.S. intervention there, "We have studied Vietnam and Lebanon and know how to get rid of Americans, by killing them so that public opinion will put an end to things."[32] Accordingly, adversaries are likely to adopt counter-intervention strategies that impose high risks of U.S. casualties.

In part because of casualty sensitivity, U.S. foreign policy also exhibits a tendency to choose military instruments such as cruise missiles or manned aircraft, which either put no U.S. personnel in harm's way or minimize the number at risk. A long-standing tenet of the "American way of war" has been a reliance on materiel over manpower, high-technology over low-tech mass.[33] The heavy reliance on the vast U.S. technological superiority, featuring in particular modern stealth and precision-guidance systems, has contributed to what Eliot Cohen has dubbed "the mystique of U.S. air power."[34] Not only do such high-technology instruments provide the necessary target discrimination to satisfy the public's demand for minimizing civilian suffering, but they also allow U.S. forces to bring massive firepower to bear without placing significant numbers of U.S. personnel in danger. The use of cruise missiles to attack suspected terrorist targets

---

as a function of the log of the casualties, although the sensitivity to casualties depended on the perceived benefits of and prospects for success. See also John Mueller, *Policy and Opinion in the Gulf War,* Chicago: University of Chicago Press, 1994, pp. 76–77, who reports empirical findings from previous conflicts to support the theory that U.S. casualties, especially under certain circumstances, erode public support for continued operations.

[32]Quoted in Barry M. Blechman and Tamara Cofman Wittes, "Defining Moment: The Threat and Use of Force in American Foreign Policy," *Political Science Quarterly,* Vol. 114, No. 1, 1999, p. 5.

[33]Russell F. Weigley, *The American Way of War,* Bloomington: Indiana University Press, 1977.

[34]Eliot A. Cohen, "The Mystique of U.S. Air Power," *Foreign Affairs,* Vol. 73, January–February 1994.

in Afghanistan in August 1998, and their threatened use against Iraqi forces in November 1998, reflected this tendency, even at the expense of predictably degraded military effectiveness.[35]

Sensitivity to Collateral Damage and Civilian Suffering

U.S. military operations are also planned with concern for minimizing collateral damage. As with American casualties, policymakers' and the public's sensitivity to collateral damage depends on the context. On the one hand, during the Vietnam conflict, many perceived that the U.S. and South Vietnamese forces were conducting indiscriminate operations—perceptions that appeared to be validated by coverage of My Lai and other actual or alleged atrocities. These perceptions combined with the indecisiveness of the war to fuel public disaffection.[36] On the other hand, there was little adverse public reaction to the hundreds of Somali deaths resulting from firefights with U.S. or UN forces, nor has there been vocal outcry since the Gulf War to Iraqi civilian deaths resulting from air strikes or economic sanctions, even though a majority of the U.S. public, at the height of the Gulf War, believed that the people of Iraq were innocent of any blame for Saddam Hussein's policies.[37] Nevertheless, significant segments of the U.S. population support minimizing risk to enemy civilians. And, even if other segments are unlikely to withdraw support as collateral damage occurs, if military planners and operators do not take substantial steps to minimize risks to civilians, general

---

[35]Paul Mann, "Strategists Question U.S. Steadfastness," *Aviation Week & Space Technology*, August 31, 1998, p. 32.

[36]As Guenter Lewy explained, "The impact of the antiwar movement was enhanced by the widely publicized charges of American atrocities and lawlessness. The inability of Washington officials to demonstrate that the Vietnam war was not in fact an indiscriminate bloodbath and did not actually kill more civilians than combatants was a significant factor in the erosion of support for the war. " Guenter Lewy, *America in Vietnam*, New York: Oxford University Press, 1978, p. 434.

[37]A *Los Angeles Times* poll (February 15–17, 1991) showed that 60 percent of respondents thought that the people of Iraq were innocent of any blame, while only 32 percent thought that the people of Iraq must share blame for Saddam Hussein's policies. Mueller, 1994, p. 316. Likewise, accidental NATO attacks on a Serbian passenger train and Kosovar refugee convoys in the early weeks of Operation Allied Force did not undermine U.S. public support for air strikes. A *USA Today* poll (April 16, 1999) taken shortly after these events showed 61 percent support (approximately the same support level as the previous week).

support will probably become less stable and, hence, potentially more vulnerable to unpredictable dips. Moreover, as with U.S. casualties, collateral damage is likely to undermine public support when combined with the perception that U.S. victory is unlikely.[38] The bottom line is that policymakers are extremely wary of authorizing actions posing high risks of significant collateral damage.

Even when the U.S. public appears willing to tolerate collateral enemy civilian injury, other members of the international community may not, and the risk of either public or international backlash is typically enough to severely constrain U.S. air operations. The added political constraints present in coalition operations are described in more detail below. But even in planning unilateral operations, the sensitivities of allies and other international actors can restrict military planning.

Often, operations must be planned with attention to minimizing enemy *combatant* casualties, in addition to minimizing injury to civilians. At the end of the Gulf War, near the Kuwaiti town of Al Jahra, allied aircraft destroyed hundreds of civilian and military vehicles that Iraqi forces were using to flee north. Reports of the carnage on the "highway of death" led General Colin Powell, the Chairman of the Joint Chiefs of Staff, to worry that the brilliant American military performance would be tarnished by images of the carnage and excessive violence against retreating forces.[39] So long as enemy forces in such situations have not signaled their surrender, they remain legally targetable.

This example illustrates how other concerns can in sometimes overlay a supplemental set of tighter constraints than international law. During the planning of Operation Deliberate Force, General Bernard Janvier, Forces Commander of the United Nations Peace Forces, ex-

---

[38]A survey by the Pew Research Center in May 1999 suggested that public support for NATO air attacks on Yugoslavia decreased because of unintended civilian casualties, combined with public concern that the attacks were ineffective. Richard Morin, "Poll Shows Most Americans Want Negotiated Settlement," *Washington Post*, May 18, 1999, p. A18.

[39]Powell's concerns are discussed in Atkinson, 1993, p. 453. It turned out that the "highway of death" air strikes destroyed many vehicles but killed few Iraqis (most abandoned their vehicles and fled into the desert). The images were more powerful than reality.

pressed concern to NATO planners regarding Bosnian Serb army casualties; targeting choices were therefore amended to reduce the likelihood that military personnel would be hit.[40] This concern over combatant casualties stemmed partly from the special considerations that drive peace enforcement operations, such as perceived impartiality (this issue is elaborated below).

RESTRICTIVE RULES OF ENGAGEMENT AND TARGETING

Political constraints emanating from concern over collateral damage have for the past several decades severely limited planning options during conflicts. During much of the Vietnam conflict, and in every military operation since, political and diplomatic pressures—especially those related to civilian damage and injury—have translated into restrictions on which targets could be struck from the air, as well as when and how.

This is not to say that U.S. forces have always operated in accordance with the law of armed conflict. Interpretations of legal obligations and factual circumstances vary. Moreover, some political pressures push against, rather than with, the humanitarian goals of the legal regime: Whereas concern for collateral damage may caution tremendous restraint in conducting air operations, concern for force protection, military effectiveness, and even financial cost may lead planners to undervalue civilian costs to operations, arguably beyond legal bounds.[41] Undeniably, though, the political factors laid out earlier restrict operational flexibility in more ways than would international law alone.

With strategic options likely to directly cause massive civilian casualties completely off the table, restrictive ROE at the tactical level are

---

[40]Ronald M. Reed, "Chariots of Fire: Rules of Engagement in Operation DELIBERATE FORCE," in Robert C. Owen, ed., *Deliberate Force: A Case Study in Effective Air Campaigning: Final Report to the Air University Balkans Air Campaign Study*, Maxwell AFB, Ala.: Air University Press, December 1999.

[41]For a critical account of U.S. targeting policy and practice in the Gulf War, see Middle East Watch, 1991. For charges of indiscriminate NATO bombing practices in Operation Allied Force, see Simon Jenkins, "NATO's Moral Morass," *The Times* (London), April 28, 1999; Mark Lawson, "Flattening a Few Broadcasters," *Guardian* (London), April 24.1999, p. 18; Fintan O'Toole, "NATO's Actions, Not Just Its Cause, Must Be Moral," *Irish Times*, April 24, 1999, p. 11.

increasingly the locus of contentious policy and legal debate. Planners often attempt to minimize collateral damage and civilian injury not only by circumscribing certain targets and conditions for engaging adversary forces but also by limiting the timing of attacks. For example, attacks on certain targets might be restricted to nighttime, when fewer persons would be expected to be in the target's vicinity.

Figure 3.1 illustrates some of the interacting constraints planners face. The figure represents a hypothetical "snapshot" of a particular crisis. The slope of the line is deliberately drawn to reflect the relatively intense political sensitivity of U.S. casualties and an implicit trade-off discounting of risks to enemy civilians.[42] Planners must select weapon systems and ROE that lie within the parameters imposed by political constraints. The three choices of ROE for the piloted platform—here, an F-16—are intentionally drawn as points along a curve to illustrate the general, although not universal, principle that efforts to reduce risk to friendly aircraft will often increase the risk of collateral damage; likewise, efforts to reduce risk of collateral damage will often place aircraft in greater danger.[43] Cruise missiles allow planners and operators to externalize most or all of the human costs of attacks by placing no U.S. personnel at risk.[44]

---

[42] *USA Today* (February 15, 1991) reported immediately following the Al Firdos bunker incident that 69 percent of the public would accept deaths of civilians near military targets in order to save U.S. lives (about three-quarters of those polled supported continued bombing in Iraqi civilian areas). Again, the willingness of policymakers and planners to trade one risk for the other will vary with contextual factor.

[43] During the planning and conduct of Operation Deliberate Force, for example, Special Instructions (SPINs) were issued to aircrews directing that (1) those attacking a bridge must make a dry pass over the target and attack on an axis perpendicular to it, releasing only one bomb per pass; (2) those carrying out suppression of enemy air defense (SEAD) strikes were not authorized without special approval to conduct preemptive or reactive strikes against surface-to-air missile sites except under certain restrictive conditions. The first of these directives was subsequently rescinded owing to concerns that it placed NATO aircrews at undue risk. Reed, 1999.

[44] W. Michael Reisman, "The Lessons of Qana," *Yale Journal of International Law*, Vol. 22, 1997, pp. 381–399, worries that such cost-externalization can skew decisionmaking toward the use of standoff weapons when the law of armed conflict would arguably demand the use of more-precise weapons (although ones that might require the attacker to accept some human risk of its own).

RAND *MR1187-3.1*

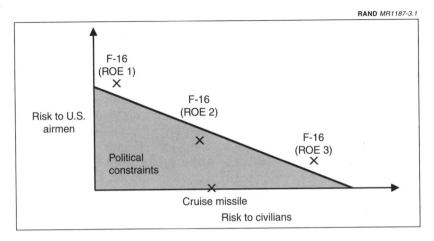

Figure 3.1—Force Protection Versus Collateral-Damage Avoidance

Extreme sensitivity to U.S. casualties may result in weapon system and ROE choices that arguably fall short of international legal obligations for collateral damage.

Note that the figure omits an important independent variable referenced briefly above: The United States may be willing to accept both a great risk to civilians and significant casualties if the stakes are high enough. As well, higher prospects for success are likely to increase political tolerance of casualty and civilian risks.

A key planning challenge is to select from among the politically, and legally, acceptable options while still achieving satisfactory levels of military effectiveness (and all within financial and resource limitations). Political concerns about friendly and civilian casualties impose ceilings of acceptable risk along the two featured axes. As higher levels of military effectiveness are demanded, the aperture of practicable options will become smaller.[45]

---

[45]However, it must be noted that effectiveness and casualty concerns are not entirely independent. For example, U.S. political leadership may be willing to tolerate higher risk levels of U.S. or civilian casualties, but only so long as they would ensure higher levels of effectiveness. And, as explained above, low levels of military effectiveness may erode public tolerance for casualties.

During the Gulf War, planners imposed strict ROE on coalition air forces, particularly when engaging urban targets: "To the degree possible and consistent with allowable risk to aircraft and aircrews, aircraft and munitions were selected so that attacks on targets within populated areas would provide the greatest possible accuracy and the least risk to civilian objects and the civilian population."[46] To this end, aircrews attacking targets in populated areas were directed not to drop munitions if they lacked positive target identification.[47] Comparable emphasis on minimizing collateral damage had generated similar restrictions on aircrews during the April 1986 bombing of Libyan terrorist-related targets: The ROE for U.S. pilots required redundant target identification checks, and, as a result, several aircraft could not release their bombs. Operation Deliberate Force ROE stated that "target planning and weapons delivery will include considerations to minimize collateral damage." Of all munitions dropped by U.S. aircraft, 98 percent were precision-guided munitions (PGMs).[48] In Kosovo operations, NATO's ROE prohibited JSTARS from passing target coordinates directly to shooters, as occurred in the 1991 Persian Gulf War.

ROE and targeting restrictions are sometimes subject to major revisions during the course of crises or conflicts. They may be modified to *expand* targeting options and operational flexibility. The Nixon administration was frustrated with unproductive air attacks on North Vietnam, which led it to remove many of the Johnson administration's limitations, particularly those that circumscribed urban areas. A similar loosening of restrictions took place during NATO's Operation Allied Force, when allied governments allowed NATO planners greater leeway to attack strategic targets after initial waves of attacks failed to move Yugoslav president Slobodan Milosevic's regime.[49]

---

[46]U.S. Department of Defense, 1992, p. 612.

[47]U.S. Department of Defense, 1992, p. 612.

[48]Reed, 1999.

[49]Michael Gordon, "NATO Air Attacks on Power Plants Pass a Threshold," *New York Times,* May 4, 1999, p. A1; Tim Butcher and Patrick Bishop, "NATO Admits Air Campaign Failed," *London Daily Telegraph,* July 22, 1999, p. 1.

However, many times ROE constrict during a campaign or operation. As operations continue, incidents or claims of excessive collateral damage can generate pressure for even tighter constraints. After the North Vietnamese accused the United States of flagrantly attacking civilian areas during December 1966 air strikes against railway targets near Hanoi, allegedly causing massive suffering, Washington responded by prohibiting attacks on all targets within 10 nautical miles of Hanoi unless specific presidential approval had been given.[50] The Al Firdos bunker incident during the Gulf War resulted in a tightening of political control over target selection; thereafter, all Baghdad targets had to be cleared beforehand with the Chairman of the Joint Chiefs.[51] This last example of extreme collateral-damage risk aversion is particularly significant: Contrary to the fears of some political and military leaders, the U.S. public's opinion of the air war was actually unmoved by the incident.[52]

One interesting phenomenon stemming from the threat of such tightening has been the recent shift of targeting decisions from political management to management by the high military command levels themselves. The military's own self-restraint is aptly demonstrated by the tight control that then–Lt Gen Michael Ryan, who commanded NATO's air forces in the Southern Region, exerted over targeting during Operation Deliberate Force. The political sensitivity surrounding the operation drove him to select personally every aimpoint, even after potential targets had already been scrubbed to avoid significant risk of civilian casualties.[53]

---

[50]Stephen T. Hosmer, *Constraints on U.S. Strategy in Third World Conflicts*, New York: Crane Russak & Co., 1987, p. 61; *Pentagon Papers*, Gravel Edition, Boston: Beacon Press, n.d., Vol. IV, p. 135.

[51]Michael Gordon and Bernard Trainor, *The Generals' War*, Boston: Little, Brown, 1994, pp. 326–327.

[52]A *USA Today* poll the following day (February 15, 1991) reported that, when asked if the shelter bombing changed their support of the war, only 14 percent responded affirmatively, while 38 percent expressed no change in their support and 41 percent said that they were more supportive of the war. Mueller (1994, pp. 317–319) also cites public opinion data showing that a majority of the public, both before and after the bunker incident, thought that the United States was making enough effort to avoid collateral damage.

[53]Reed, 1999. As a related example, in April 1999, after a U.S. warplane mistakenly hit a refugee convoy, procedures were modified to require that American aircrews over Kosovo radio for authorization before striking military convoys. Elaine Harden and

THE ASYMMETRY OF CONSTRAINTS

These constraints would be challenging even if both sides felt their effect equally. In practice, however, it is often the case that the United States and its allies operate under these constraints while their adversaries do not. This section explores the implications of this asymmetry.

Adversaries will typically be less constrained than the United States and its allies by international legal norms. The United States generally benefits from stability and international order; its adversaries are often interested in overturning that order: "Since law is generally a conservative force, it is more likely to be observed by those more content with their lot."[54] Apart from possible differences in commitment to international norms and preservation of international law in general, some adversaries are likely to view the United States, with its vastly superior military technology, as a manipulator of the law of armed conflict for its own benefit.

Strategic setting is critical to this analysis: What is a small-scale contingency for the United States may be a major war for an adversary. Conflict with the United States may implicate an adversary state's or regime's most vital interests and may strain its willingness to remain bound by international legal rules that, at a given time, may favor U.S. military dominance.

Uneven adherence to the law of armed conflict between the United States and adversaries provides these adversaries with manifold opportunities for strategic and tactical exploitation. Adversaries often expect that U.S. political resolve will erode as collateral damage, civilian injury, or U.S. casualties mount, especially when the most vital U.S. interests are not at stake or allied support is shaky.[55]

Opportunities for exploiting constraints on U.S. operations—opportunities that adversaries have historically seized with some success—

John M. Broder, "Clinton's War Aims: Win the War, Keep the U.S. Voters Content," *New York Times*, May 22, 1999, pp. A1, A6.

[54]Louis Henkin, *How Nations Behave: Law and Foreign Policy*, New York: Praeger, 1968, p. 49.

[55]Daniel Byman and Matthew Waxman, "Defeating US Coercion," *Survival*, Vol. 41, No. 2, Summer 1999.

expand in the urban environment. Knowing that U.S. planners and operators are obliged to verify their target objectives, adversaries can disperse dual-use sites, camouflage military assets, and otherwise hinder U.S. information-gathering. Knowing that U.S. planners and operators will avoid incidental civilian losses, adversaries can commingle military and civilian assets and persons. And knowing that U.S. planners and operators will avoid attacks likely to cause excessive civilian damage, adversaries can manipulate the media following attacks to portray exaggerated destruction.

In adopting these techniques, adversaries hope that the potential for U.S. casualties or political backlash resulting from anticipated collateral damage will deter U.S. intervention. If the United States intervenes, these techniques aim to confront U.S. planners with a dilemma: refrain from attacking (or attack under extremely tight operational restrictions) certain targets, therefore risking degraded military effectiveness, or attack those targets effectively and risk collateral damage or perhaps higher levels of U.S. casualties.

An adversary's ability to exploit constraints on U.S. operations depends on a number of factors, including the adversary's own bases of support, its strategy, and its propaganda capabilities. Autocratic, dictatorial regimes typically maintain tight control over the media. While manipulating the content of information flowing to its own population, these regimes can also influence the timing and, indirectly, the substance of information disseminated abroad by selectively permitting journalistic inspection.

The Viet Cong and North Vietnamese were both notoriously obstructive and invitingly supportive of Western television, depending on the situation. Yugoslav President Slobodan Milosevic displayed a similar pattern of cracking down on independent media each time crises flared with the international community.[56] During NATO's Operation Allied Force, Milosevic shut down independent newspapers and radio stations inside Yugoslavia, used state-run television to stoke nationalist reactions, electronically jammed some U.S. and NATO broadcasts intended for the Serbian populace, and prohibited

---

[56]Chris Bird, "Kosovo Crisis: Yugoslav Media Fear Crackdown Amid War Fever," *Guardian*, October 8, 1998, p. 15; Jane Perlez, "Serbia Shuts 2 More Papers, Saying They Created Panic," *New York Times*, October 15, 1998, p. A6.

the Western press from much of Kosovo (while granting it permission to film bombed sites, especially in major cities such as Belgrade and Novi Sad).

To be sure, the efforts of an adversary to profit from civilian casualties often fail and may even prove counterproductive if the American and international public views the adversary leadership as being at fault. But even when adversary efforts to exploit collateral damage do not result in a tightening of the United States' self-imposed constraints, they publicly put U.S. policymakers on the defensive and may harden the resolve of adversaries who expect American political will to dissolve.

The characteristics of urban environments discussed earlier—population density, the proximity of civilian and military targets, and shared civilian-military assets—provide adversaries with many opportunities to exploit asymmetrical constraints. The potential for large civilian death or injury tolls, the ease of situating military assets near or camouflaging them among civilian assets, and the intense media scrutiny surrounding incidents of collateral damage facilitate adversary shielding tactics. Evidence from recent conflicts demonstrates the tendency of adversaries to employ such tactics, frequently with some success.

Adversaries often deliberately commingle civilian and military assets or persons in an effort to shield them from attack. In Somalia, U.S. and UN forces frequently encountered hostile militiamen firing from behind women and children. U.S. forces trying to aim at armed threats from the air found that militiamen took advantage of crowded streets to open fire and then disperse or blend into crowds of civilians.[57]

Using civilian assets or persons to shield military targets is especially easy in urban environments, where civilian objects and persons dramatically increase the risk of collateral damage in any attack from the air. As alluded to above, North Vietnamese forces routinely capi-

---

[57]Bowden, 1999, p. 46, describes numerous examples. In one incident, U.S. forces encountered "[a] Somali with a gun lying prone on the street between two kneeling women. The shooter had the barrel of his weapon between the women's legs, and there were four children actually *sitting* on him. He was completely shielded in noncombatants, taking full cynical advantage of the Americans' decency."

talized on public U.S. declarations restricting attacks in densely populated areas by storing military supplies in such places. During the Gulf War and in subsequent U.S. air operations against Iraq, the Iraqi government refused to evacuate civilians known to be situated close to key targets in Baghdad and other cities;[58] according to the Defense Department's post-war account, "[p]ronouncements that Coalition air forces would not attack populated areas increased Iraqi movement of military objects into populated areas in Iraq and Kuwait to shield them from attack."[59]

The potential to exploit vertical proximity of civilians and military objectives in urban environments can be seen in Palestine Liberation Organization (PLO) practices during the 1982 Israeli incursion into Lebanon. Contravening its legal obligations to segregate the civilian population from military objectives, PLO forces in towns and cities placed artillery and anti-aircraft weapons on top of hospitals and religious buildings, in an effort to negate the technological superiority of the Israeli Defense Forces and Israeli Air Force. Upon retreating to Beirut, some PLO forces allegedly positioned themselves and their military equipment in lower floors of high-rise apartment buildings and forced civilian tenants to remain in upper floors. Civilian injury tolls were substantial, although Israeli forces' strict ROE often resulted in successful shielding of legitimate PLO military targets.[60]

Adversaries also routinely take advantage of the special protected status accorded certain types of structures, such as medical or cultural buildings. Members of the PDF used Santo Tomás Hospital for sniper activity in attempting to repel U.S. forces during Operation Just Cause.[61] A cache of Iraqi Silkworm surface-to-surface missiles were discovered inside a school in a densely populated Kuwait City

---

[58]Indeed, Saddam Hussein has used his authoritarian state apparatus with great success to put civilians in harm's way when faced with threats of air strikes. Barbara Crossette, "Civilians Will Be in Harm's Way If Baghdad Is Hit," *New York Times*, January 28, 1998, p. A6.

[59]U.S. Department of Defense, 1992, p. 615.

[60]Parks, 1990, pp. 165–166. A far more critical account of the Israeli Air Force's bombing operations is found in Martin van Creveld, *The Sword and the Olive: A Critical History of the Israeli Defense Force*, New York: Public Affairs, 1998, p. 297.

[61]Americas Watch, 1990, p. 26.

area,[62] and Iraq positioned two fighter aircraft adjacent to the ancient temple of Ur during the Gulf War.[63] During Operation Allied Force, the Yugoslav armed forces reportedly used churches, schools, and hospitals to shield troops and equipment against NATO air strikes, knowing that NATO forces operated under tight ROE and that, even if Serbian practices justified attacks on these targets, NATO planners were eager to comply with international legal restrictions and to avoid potential political fallout from destruction of these sites.[64]

The use of civilian structures, including those with special cultural significance, to shield military targets stems not only from a willingness by some adversaries to breach international norms but also from asymmetries in the costs each side associates with the demolition of those structures. The potential effectiveness of adversary shielding techniques is therefore highly context-dependent. U.S. and Republic of Korea (ROK) forces attempting to dislodge invading North Korean forces from Seoul would likely be far less willing to demolish civilian property than if they were attempting to capture Pyongyang. However, the United States and ROK would probably do so in Seoul if required; the willingness to cause (and in this case, sustain) civilian destruction is partly a product of military necessity.

In MOOTW, such as efforts to maintain order or separate local combatants, strategic demands on planners may place premium costs on destroying civilian property if doing so would inflame local popular resentment. In each of the above cases, the potential efficacy of shielding depends on the relative costs of civilian damage that each side must internalize, as well as the relative commitments of each side to international legal obligations.

As pointed out earlier, human-shield tactics may backfire, particularly if viewed locally or abroad as barbaric. But some adversaries seem willing to bear that risk in the face of otherwise overwhelming U.S. military might.

---

[62] U.S. Department of Defense, 1992, p. 613

[63] U.S. Department of Defense, 1992, p. 615.

[64] Elaine Harden and Steven Lee Myers, "Bombing United Serb Army As It Debilitates Economy; Yugoslav Rift Heals, NATO Admits," *New York Times*, April 30, 1999, pp. A1, A13.

Of the various potential U.S. adversaries, irregular forces are most able to exploit asymmetric constraints.[65] Adherence to the principles of target discrimination becomes much more difficult when there are few, if any, physical markings to distinguish combatants from noncombatants. Moreover, some irregular military organizations may have little or no incentive to adhere to international norms and are, therefore, even more likely to capitalize on the United States' self-imposed constraints. Testifying to the extent to which adversaries will likely go, some PDF units were trained before Operation Just Cause to disperse, dispose of their uniforms in favor of civilian clothes, and return to Panama City to repel any U.S. intervention or invasion.

Blurred distinctions between combatants and noncombatants complicate target discrimination and facilitate human-shield tactics.[66] For example, in Somalia and southern Lebanon, the UN and Israel, respectively, faced enemy personnel virtually indistinguishable from the heavily armed civilian populace. This fact alone complicates targeting, especially from the air. It also allows enemy forces to blend into civilian crowds, taking advantage of attacking forces' restrictive ROE or forcing them to risk hitting civilians.[67]

---

[65]*Irregular forces* here refers to guerrilla and militia units and other adversary forces lacking official uniforms and other insignia used to differentiate combatants from noncombatants.

[66]It is in part because of the difficulties of applying traditional international legal principles to guerrilla and irregular force contexts that the legal regime sometimes contains different provisions for internal, as opposed to international, armed conflicts. Almost any U.S. operations will involve application of international armed conflict law; this report does not discuss legal issues specific to internal conflicts.

[67]The law of armed conflict attempts to regulate these practices, although with little success in balancing the exigencies of counterguerrilla operations with civilian protection. Article 44(3) of Protocol I, for example, states that:

In order to promote the protection of the civilian population from the effects of hostilities, combatants are obliged to distinguish themselves from the civilian population while they are engaged in an attack or in a military operation preparatory to an attack. Recognizing, however, that there are situations in armed conflicts where, owing to the nature of the hostilities an armed combatant cannot so distinguish himself, he shall retain his status as a combatant, provided that, in such situations, he carries his arms openly:

1) during each military engagement, and

Calculating proportionate military responses is especially vexing against irregular forces, because the blurred distinction between armed foes and civilian bystanders confuses determinations of threats. During a September 1993 ambush of UN forces by Somali militiamen using women and children as shields, U.S. Cobra helicopters shot into the crowd. Italy and other coalition members protested vehemently that the U.S. response was excessive, to which Major David Stockwell, the UN military spokesman, replied: "In an ambush there are no sidelines for spectators."[68]

The Somalia case also illustrates that nonstate military organizations often have tremendous ability to manipulate domestic and international public opinion, even when they lack monopoly control over state infrastructure. Aideed garnered support both within and outside Somalia by exploiting civilian casualties resulting from engagements with UN forces (many of them attributable in part to Aideed's deliberate use of civilian crowds to shield his militia personnel), despite the fact that Somalia lacked high-technology communications systems for disseminating propaganda.[69]

CONCLUSION

Legal norms and political pressures will constrain all U.S. military operations. Experience since the Vietnam War teaches that competing concerns of force protection, collateral damage, and other political issues can severely restrict operational flexibility. Air planners face the daunting task of designing strategically effective operations under pressures and duties that partially negate USAF capabilities. But just as policymakers need to understand how tight restrictions they may impose on tactical and operational choices may reduce military potency, military planners need to appreciate that satisfying political and diplomatic demands may be vital to sustained support

---

2) during such time as he is visible to the adversary while he is engaged in a military deployment preceding the launching of an attack in which he is to participate.

[68]Leslie Crawford, "Unrepentant Peacekeepers Will Fire on Somali Human Shields," *Financial Times*, September 11, 1993, p. 4.

[69]James O. Tubbs, *Beyond Gunboat Diplomacy: Forceful Applications of Airpower in Peace Enforcement Operations*, Maxwell AFB, Ala.: Air University Press, 1997, p. 35.

for military operations. In other words, the same restrictions that an operator views problematically as "constraining" may be critical enablers of military options at the highest strategic levels.

These legal and political constraints are tightest, and their effects most magnified, in urban environments. While striving to keep U.S. forces out of harm's way and design operations sufficiently capable of achieving strategic objectives, planners and operators must avoid unintended civilian injury. Because urban environments are characterized by dense populations and collocated or shared civilian-military assets, the range of available options that satisfy these competing objectives will often be narrow.

Lacking an equivalent degree of commitment to international norms and facing very different political, diplomatic, and strategic exigencies than the United States, adversaries are likely to exploit the asymmetrical constraints to the maximum extent possible. Adversaries operating in urban settings will have tremendous incentive to breach their own legal obligations, hoping to capitalize on the propaganda effects of collateral damage. Furthermore, urban environments provide adversaries with convenient means to do so; the features of urban environments that already pose problems for U.S. planners also facilitate deliberate mingling or camouflaging of civilian and military targets in an effort either to shield military assets from attack or to increase the potential political or resource costs to the United States of hitting them.

An appreciation of these political constraints—from both the U.S. planner's and operator's perspective and from the adversary's—is critical to designing USAF concepts of operation. The unique capabilities of U.S. air forces, enhanced by continued technological advances in key areas, will give the USAF a key role in future urban operations across the spectrum of conflict. The USAF's contributions will be maximized by tailoring its operational concepts around the legal and political context in which its missions will arise.

Having explored the legal and political constraints associated with urban air operations, we now turn, in Chapter Four, to the challenges associated with detecting and striking targets in the urban *physical* environment.

Chapter Four

AEROSPACE OPERATIONS AND THE URBAN PHYSICAL ENVIRONMENT

In the preceding chapters, we have placed urban military operations in a broader strategic, political, and legal context. With those issues framing our analysis, in the next three chapters we turn to the operational, tactical, and technical challenges facing airmen in the urban physical environment.

INTRODUCTION

To examine how urban terrain affects the requirements for demanding missions such as identifying, tracking, and targeting individual people or small groups in streets, on rooftops, or in rooms in a complex urban environment, we have divided this chapter into three main sections. The first section presents general observations on the nature of urban topography and shows how important physical characteristics of the urban environment, such as street width, building height, and wall thickness, can be used to divide a city into militarily useful Urban Terrain Zones (UTZs). It presents also some data on the proportion of overall urban areas that fall into each UTZ. The second section examines the challenges that different UTZs present for airborne surveillance and reconnaissance and outlines how those challenges might be met. The third section looks at the difficulties associated with employing aerial weapons on urban terrain. The final section deals with airlift in urban environments.

THE NATURE OF URBAN TERRAIN

A number of characteristics of urban terrain make it unique from a military perspective. First, and most obvious, are the man-made buildings, streets, and other structures that dominate urban terrain. These structures vary widely in size, height, type of construction, wall thickness, number and size of windows and doors, etc. However, they share certain basic characteristics that give urban terrain its most distinctive macro features.

The most striking differences between urban terrain and most natural types of terrain are the verticality and regularity of man-made structures. The following passage from a study that Richard Ellefsen led for the U.S. Navy emphasizes these differences:

> What are (the) shapes and forms found in the city and how are they similar to or different from forms found in non-urban settings? Some broad similarities occur since cities, after all, are built upon a segment of natural terrain. Relief is a factor, though modified (usually reduced) by cutting and filling. Drainage must be provided (in modified form by streets and drains). Land and water interfaces are present but are often sharpened by levees and dredging. Weather and climatic extremes are modified through the absorption and re-radiation of heat by building and street surfaces. Residential landscaping also reduces extremes. Precipitation may be increased by the injection of more hygroscopic nuclei into the atmosphere over a city and wind patterns are altered by buildings and street "canyons."
>
> Special shapes and patterns result from the creation of cities. Planimetric shapes, especially the regularity of a grid street pattern, have virtually no counterpart in nature; some geologically dictated drainage patterns vie to only a small degree. But, it is in its profile that the city differs most markedly from natural forms. The cluster of towers of a city's core is indeed symbolic of the traditional conceptual view of a city. The most significant contribution to its non-natural character is the visual impact of the contrast between the horizontality of streets and building tops and the verticality of the building walls. Natural terrain counterparts, e.g. Devil's Tower or the Grand Canyon are few and rate special recognition.[1]

---

[1]Richard Ellefsen, Bruce Coffland, and Gary Orr, *Urban Building Characteristics: Setting and Structure of Building Types in Selected World Cities*, Dahlgren, Va.: Naval

The unique character of urban terrain dramatically restricts line of sight (LOS). Imagine an observer standing on a sidewalk in the middle of a block on a straight street bordered on both sides by three-story row houses. The observer can see across the 15–25-meter-wide street and a great distance down the street. However, the observer, who is effectively at the bottom of a shallow "urban canyon," cannot see what is on the other side of the facing row of buildings. Now imagine there is a tall pole next to our observer and the observer begins to climb it. Not until the observer climbs to a height greater than the surrounding buildings do the roofs of buildings across the next street over become visible, then the upper floors, and finally the entire building faces. If the observer climbs higher still, the sidewalk and then the next street over itself will become visible.

How high the observer will have to climb to see over intervening buildings will depend on two characteristics of the urban terrain in the area: the tallness of the buildings and the distance between the buildings. The taller the surrounding buildings are, the higher the observer will have to climb to see over them. Similarly, the shorter the distance between the buildings, the higher the observer will have to climb to see over them. Figure 4.1 illustrates these relationships.

The illustration in Figure 4.1a shows that the observer must be highest to see over a tall building and across a narrow street. Figure 4.1b shows that the observer can be much lower to look over the same building but across a wider street. The observer can be at a similar height to see over a shorter building on a narrow street, as illustrated in Figure 4.1c. Finally, the observer can be quite low indeed to look over a short building on a wide street, as shown in Figure 4.1d.

Note the middle building in all four views. No matter how high the observer may climb, the right-hand face of that building will not be in view, nor will that of the building on the right. In other words, it is impossible for any single observer to have LOS to all of the building faces in a city at any one time. This is an important point that we will return to in the surveillance and reconnaissance discussion in the next section. First, we turn our attention to quantifying and classifying the aspects of urban form discussed above.

Surface Weapons Center (now the Naval Surface Warfare Center), Dahlgren Laboratory, 1977, p. 2.

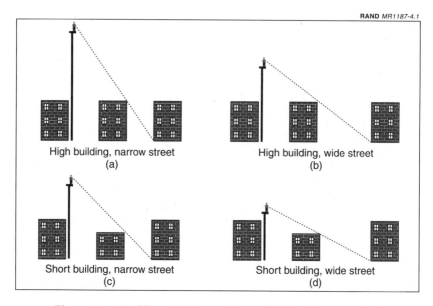

RAND *MR1187-4.1*

Figure 4.1—Building Height and Street Width Affect Required
Observer Height

Urban Terrain Zones and Their Characteristics

Most readers will have seen aerial photographs of large urban areas, such as Los Angeles or New York, or the cities themselves from the air while flying in commercial, military, or private aircraft. The initial impression these images provide is one of a teeming jumble of buildings of all shapes and sizes. A closer inspection (or further recollection) would show that, in fact, buildings and streets in different parts of the city are different shapes and sizes but tend to be similar in shape and size to their neighbors.

This should not be surprising, since the same economic considerations that lead a property owner to build a ten-story apartment building will lead another to do the same. The same holds true for commercial and industrial districts of a city. Furthermore, at a certain time, economic development tends to be concentrated in particular districts, so that many of the buildings in a particular area will have been constructed at roughly the same time with similar materials and techniques. In addition, zoning laws in developed countries tend to reinforce these economic factors by grouping buildings according to function (residential, industrial, commercial, etc.).

These systematic differences between, and similarity within, different parts of a city make it possible to classify urban terrain into different zones according to the physical characteristics of the buildings and streets in a given area.

During the late 1970s and early 1980s, Dr. Richard Ellefsen, a geography professor at San Jose State University working as a consultant for the Naval Surface Weapons Center (now the Naval Surface Warfare Center) and Aberdeen Proving Ground, began to develop a militarily useful urban terrain classification system based on the physical characteristics and spatial patterns found in different parts of cities. He based his Urban Terrain Zones (UTZs) on extensive data on the size, height, type of construction, separation, etc. of buildings in samples of cities throughout the world. The city data were generated from painstaking examination of aerial photographs. Measurements and estimates were confirmed by visiting each city and conducting a ground survey of small portions of various areas. Obviously, to develop the detailed data required to accurately divide cities into UTZs requires a great deal of human effort, making it too expensive and time-consuming to select a large, random sample of cities for analysis. Instead, Dr. Ellefsen used his judgment to develop a strategy for selecting cities in each of his studies.[2] He made sure the cities varied on the following dimensions: population, latitude and climate type, terrain character (relief), type of port services, importance as an administrative center, and evolution and development.

Given the carefully stated criteria Dr. Ellefsen used in selecting his samples, it is probably safe to assume that conclusions based on the characteristics of these cities are fairly robust and accurately reflect the physical characteristics of most urban areas around the world.

---

[2]In theory, a large random sample of cities would yield data that more closely reflect the true mix of physical characteristics and spatial patterns found in all of the world's cities. However, the resources and time required to produce accurate aerial photographs, analyze and classify different areas within the cities, and confirm measurements are so large that the random-sample approach is probably not practical.

Our confidence in the validity of conclusions based on these purposive samples was increased by two factors. First, people in cities everywhere respond in similar ways to the universal economic incentives mentioned above. Second, since the 1960s, there has been an increasing convergence, or universality, of construction techniques and materials throughout the world.[3] So, to the extent that there is less variation in the physical form of cities than in the past, generalizations based on purposive samples will tend to be more reliable.

In his most recent work (completed in 1999), Dr. Ellefsen has collapsed his original 16 UTZs into seven new classifications. Table 4.1 lists the new UTZ classifications and their typical location within a city.

In this latest study, the following 14 cities were used in the purposive sample: Helsinki, Finland; Braunschweig, Germany; Stuttgart, Germany; Vienna, Austria; Uppsala, Sweden; Salzgitter, Germany;

Table 4.1

Urban Terrain Zone Classification System

Urban Terrain Zone (UTZ)		Typical Location Within City
I	Attached and Closely Spaced Inner-City Buildings	City Core
II	Widely Spaced High-Rise Office Buildings	City Core and Edge of Built-Up City (e.g., near airports)
III	Attached Houses	Near City Core
IV	Closely Spaced Industrial/ Storage Buildings	Along Railroads Near Core and on Docks
V	Widely Spaced Apartment Buildings	Edge of City
VI	Detached Houses	Near Core and in Suburbs
VII	Widely Spaced Industrial/ Storage Buildings	At City Edge Near Highways

SOURCE: Adapted from Richard Ellefsen, *Current Assessment of Building Construction Types in Worldwide Example Cities*, Report prepared for Naval Surface Warfare Center, Dahlgren Laboratory, Dahlgren, Va., 1999, p. 16.

---

[3]For more on emerging universal building-construction techniques, see Ellefsen, Coffland, and Orr, 1977, Chapters 2 and 3; and Richard Ellefsen, *Current Assessment of Building Construction Types in Worldwide Example Cities*, Report prepared for Naval Surface Warfare Center, Dahlgren Laboratory, Dahlgren, Va., 1999.

Bremen, Germany; Tel Aviv, Israel; Tunis, Tunisia; Colombo, Sri Lanka; Kuala Lumpur, Malaysia; San Jose, Costa Rica; Panama City, Panama; and Caracas, Venezuela. Figure 4.2 shows the overall proportion of the total area of all cities that fell into each UTZ.

Figure 4.2 makes it clear that the cities Dr. Ellefsen studied are made up primarily of detached houses and widely spaced apartment buildings. Together, these two UTZ types (V and VI) account for more than 60 percent of the total area of the cities studied. This is not the mental image of "city" most of us carry in our heads. In fact, the word *city* is more likely to evoke images of attached and closely spaced buildings (UTZ I) or widely spaced high-rise office buildings (UTZ II), UTZs that cover only about 3 percent and 1 percent, respectively, of the land area of the cities studied. However, although accounting for only a small fraction of total city land area, these UTZs contain the most valuable land and often the most important cultural, economic, and administrative structures in a city. This fact

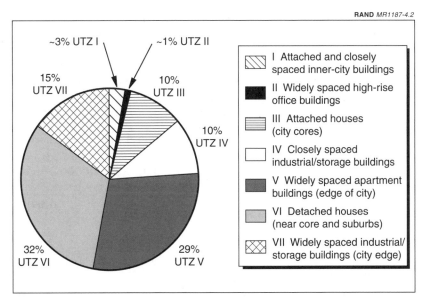

RAND *MR1187-4.2*

SOURCE: Adapted from Richard Ellefsen, *Current Assessment of Building Construction Types in Worldwide Example Cities*, Report prepared for Naval Surface Warfare Center, Dahlgren Laboratory, Dahlgren, Va., 1999, p. 38.

Figure 4.2—Proportion of Surveyed Cities in Each UTZ

gives these small zones a practical and symbolic importance far out of proportion to the area they cover. Therefore, they are likely to be the focus, or even the objective, of both U.S. and adversary actions during urban operations. It is also possible that an adversary could exploit these city-core UTZs, with their extremely restrictive LOS characteristics (explained in detail below), to pose maximum difficulties for U.S. air operations.[4]

In addition to determining the fraction of city area that fell into each UTZ, Dr. Ellefsen measured the following building attributes for each UTZ type: average building footprint, average building height, predominant construction type, predominant wall-construction material, average wall thickness, pitched or flat roof, approximate venting (doors and windows) in outer walls, and average separation between buildings. Table 4.2 summarizes his values for these building attributes.

Building heights and separations of each UTZ,[5] as well as other building attributes, are determined by the prevailing land values, construction techniques, modes of transportation, and intended function of the buildings at the time of their construction. Therefore, warehouse districts have different average building heights, wall thicknesses, and building separations from downtown commercial districts; likewise, residential areas near old city cores built at a time when trolleys and foot traffic were the dominant modes of transportation are very different from modern, automobile-age suburbs.

The discussion in the previous section indicated that average building height and average building separation will be crucial factors in determining how much of a city airborne platforms can potentially see from a given altitude. Combining the data in Figure 4.2 with those in Table 4.2 reveals that, for the cities studied, the median

---

[4]The importance and likelihood of city-core operations, combined with the limited ability of aerial platforms to effectively monitor these areas (explained below), makes the development of the sort of distributed Unattended Ground Sensor (UGS)–based urban sensor network described later in this chapter especially critical for effective air operations in the urban core.

[5]Unless otherwise noted, *street width* is used as shorthand for building separation in this chapter. It includes sidewalks in city centers and yards, sidewalks, open space around buildings, as well as streets in suburban and industrial areas.

Table 4.2

Building Attributes by Urban Terrain Zone

UTZ Type	Footprint (m²)	Avg. Height (m)	Construction Type	Wall Material	Wall Thickness (cm)	Roof	Approx. Venting (%)	Avg. Separation (m)
I	2,000	30	Mass	Brick	30	Flat	70	5
II	4,000	45	Frame	Glass	10	Flat	90	30
III	500	6	Mass	Brick	20	Both	15	20
IV	4,000	9	Both	Brick	25	Both	5	25
V	1,000	20	Frame	Block	20	Flat	70	50
VI	100	6	Mass	Block	25	Pitch	25	15
VII	10,000	9	Frame	Corrugated Steel	1	Both	5	70

SOURCE: Adapted from Richard Ellefsen, *Current Assessment of Building Construction Types in Worldwide Example Cities*, Report prepared for Naval Surface Warfare Center, Dahlgren Laboratory, Dahlgren, Va., 1999, p. 38. Mass construction usually uses stone or brick and relies on thick (up to 30-inch) load-bearing walls to support both structural and dynamic loads. Frame construction uses a wood, steel, or reinforced-concrete "skeleton" with thin walls that bear no loads.

building height is 9 meters (9 m) and the median street width is 50 m. However, these two values do not often occur together. The 9-m median building height is for UTZ IV, closely spaced industrial/storage buildings, which have an average separation of only 25 m, not the median 50 m. Similarly, the median separation of 50 m is associated with UTZ V structures, widely spaced apartment buildings, with an average height of 20 m rather than the median 9 m. The unique combinations of building height and street width within each UTZ make it impossible to use average values in analyzing LOS within an urban environment.

The values for each individual UTZ must be used to achieve accurate results. It is worth noting that the vast majority of buildings—68 percent of the total in UTZs III, IV, VI, and VII—are, on average, 9 m (about 3 stories) or less in height. Only about 4 percent—in UTZs I and II—average over 30 m, or about 10 stories, in height.

With the preceding discussion of urban form as background, and before moving on to a discussion of surveillance and reconnaissance issues, it is worthwhile to take a look at how urban terrain changes the ground-based threat to air operations. The next subsection discusses some of the more salient aspects of urban air defenses.

Air Defenses on Urban Terrain

Our analysis focuses primarily on more-limited operations, ranging from peace operations to NEOs. In these situations, the most common threat to aerospace operations stems from manportable air defense systems (MANPADS), small arms, and smaller mobile surface-to-air-missiles (SAMs; such as SA-8s). Of these, the MANPADS threat is the most worrisome.

While it might be possible to deploy some small radar-guided SAM systems, such as the SA-8 or Crotale, within urban areas, in many instances the presence of intense near-in clutter and buildings blocking the radar's field of regard would limit those systems to a fraction of their normal capability.

Anti-aircraft artillery 20mm or larger would also be difficult for U.S. adversaries to use effectively in more limited conflicts. These weapons are very large, must be towed or vehicle-mounted, require several crewmembers, and could not take full advantage of their range unless emplaced in clear areas or on top of buildings—characteristics that would make them easy to detect and attack before they could cause much trouble. Therefore, the primary threats facing U.S. air assets operating over urban terrain are likely to be shoulder-launched SAMs, heavy machine guns, small arms, and other infantry weapons.

MANPADS. While large numbers of small arms and machine guns can be lethal below 3,000 feet (ft), the shoulder-launched man-portable air defense systems (MANPADS) pose the greatest threat. These small missiles, usually with infrared guidance, can be carried, targeted, and launched by a single person. Older systems, such as the 1960s-vintage Russian SA-7, are available in large numbers throughout most of the world. They are effective against helicopters and fixed-wing aircraft operating at altitudes as high as 10,000 to 15,000 ft and at ranges between 3 and 4 nautical miles (nmi). Over the past decade, newer, more-capable systems manufactured in Europe and the Far East, such as the French Mistral and Chinese QW-1, have been widely exported. If this trend continues, more-modern systems—with maximum effective altitudes up to 20,000 ft and vastly improved guidance systems—could become increasingly common throughout the world within a decade.

However, MANPADS are not without shortcomings. MANPADS are difficult to detect and/or attack prior to launch, because they rely on visual (as opposed to radar-guided) target detection. But this very mode of target acquisition, which relies on a single operator, sometimes assisted by a spotter, often causes MANPADS to miss target-engagement opportunities, either because operators do not see the targets while they are in range or because targets are acquired while in range but fly out of range before they can be engaged. Adversaries operating in intact cities could take advantage of emerging wireless data-transfer technologies, such as cellular phones or wireless modems linked to laptop computers, to enable a number of MANPADS teams to share target information. Alternatively, they could link a network of observers with one or more MANPADS teams.

Such a network would greatly increase MANPADS effectiveness by providing early warning of approaching targets and by allowing targets detected by one team, or observers, to be engaged by another team in a better firing position. In addition, it could make the MANPADS teams even harder to detect and attack by allowing them to remain hidden on the upper floors of buildings until a target, tracked by unarmed observers around the city, entered their engagement envelope. Using advance knowledge of target location, speed, altitude, and heading, they could rapidly acquire the target and launch their missile, then disappear back inside.

MANPADS networks such as this would pose a serious threat to many USAF platforms, such as AC-130s and Predator—currently the most useful platforms in urban operations—particularly if these platforms attempted to operate at altitudes between 5,000 and 10,000 ft. At these altitudes, their sensor and weapon systems are most accurate and effective, but the platforms are also clearly visible and well within the engagement envelope of even the oldest MANPADS. Helicopters would also be extremely vulnerable to a MANPADS network such as the one described above.[6]

These considerations do not mean that the low-to-medium-altitude regime will be denied to U.S. forces in every future urban military operation, particularly if advances in infrared countermeasures (IRCMs) can spoof or damage IR seekers. However, at least some of the time, these altitudes will be dangerous for manned aircraft, and it may be necessary to perform reconnaissance and surveillance using a combination of high-altitude aircraft, such as Global Hawk, and small, inexpensive unmanned platforms, such as low-altitude mini-UAVs, micro–aerial vehicles (MAVs), or unattended sensors.[7]

The next section looks at individual UTZ average building height and separation in greater detail as it explores the geometry of urban surveillance and reconnaissance. It emphasizes using small, survivable UAVs, unattended ground sensors (UGS), and other methods that do not require manned platforms to be put at risk when exposed

---

[6]This situation highlights the need for the development and deployment of advanced infrared countermeasures (IRCM) to increase the survivability of aerial platforms operating over urban terrain.

[7]Research conducted by Randall Steeb and others at RAND suggests that deploying low-cost UAV decoys (6 decoys per real surveillance UAV) can significantly increase the survivability of the UAV surveillance platforms in high-threat air defense environments.

to the type of MANPADS threat just discussed. The final section of this chapter deals with the challenge of urban weapons delivery. It is concerned with venting (seeing and shooting through openings such as windows and doors), as well as building height and street width.

SURVEILLANCE AND RECONNAISSANCE IN THE URBAN ENVIRONMENT

The alternating vertical and horizontal surfaces that dominate as the features of urban terrain present unique and difficult challenges for surveillance and reconnaissance. Unlike open, rural spaces such as rolling farmland, prairie, or desert, buildings on urban terrain block sight lines so that large sensor platforms that require a clear LOS to their targets (such as JSTARS) cannot effectively sweep huge areas, looking for moving vehicles and other targets on the ground. In addition, in many situations, day-to-day activity continues in the city and provides a dense clutter of moving people, vehicles, and electromagnetic signals that can be exploited as cover by paramilitary forces opposed to U.S. interests or operations. During Operation Allied Force, for example, Serb army units often used the mass of Kosovar Albanian refugees moving on the province's rural roads to help mask their movement, mingling their convoys with refugees. Several times, NATO aircraft bombed these intermingled groups of vehicles, causing substantial civilian casualties. Although occurring in rural settings, these incidents underscore the need for high-resolution sensors or ground observers to positively identify targets in environments where "bad actors"—whether terrorists, guerrillas, paramilitary groups, or conventional military units—are likely to be mingled with civilians or friendly troops.

Urban surveillance and reconnaissance are more demanding than surveillance and reconnaissance on most other types of terrain for two reasons. First, poor LOS and intense clutter make large cueing sensor platforms that work so well in more-open environments of much less value in an urban setting. Second, the potential for intermingled civilians, friendly forces, and adversaries—especially, but not exclusively, in limited urban operations—makes it critical to clearly identify all targets with high-resolution sensors or ground observers before attacking targets from the air. These unique requirements call for different sorts of surveillance and reconnaissance systems and techniques from those developed to conduct mobile armored warfare. The next two subsections discuss the urban cueing and target-identification/-tracking problems.

Sensor Cueing in the Urban Environment and a New Cueing Concept

Traditional Cueing. Imagine that we have a surveillance aircraft equipped with a sophisticated Moving Target Indicator (MTI)/synthetic aperture radar (SAR) and computers that can detect, track, and classify moving ground vehicles. This aircraft normally operates at altitudes of 30,000 to 35,000 ft. Assume that, on flat terrain, this aircraft can detect moving vehicles 100 nmi away. The maximum depression angle of the aircraft's radar beam is 45°. Our notional surveillance platform can sweep a sector 120° in azimuth. Figure 4.3 illustrates how such a platform would perform over fairly flat and open terrain.

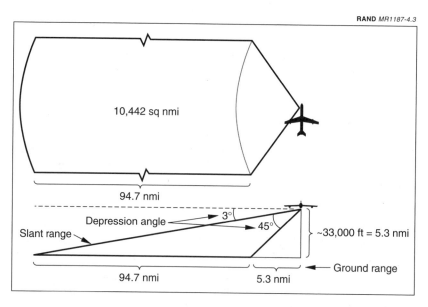

Figure 4.3—Surveillance/Reconnaissance Platform Coverage on Flat, Open Terrain

The sensor platform can detect vehicles in a huge area, almost 10,500 square nautical miles, because there are few obstructions to block the radar signals from reaching targets out to the system's maximum range of 100 nmi. At this range, the depression angle is approximately 3°.[8]

Now suppose that we want to use this aircraft to monitor vehicle traffic in an urban area. We will immediately have to face some unpleasant facts. First, the maximum range of our platform will be dramatically reduced. For example, if we want our aircraft to monitor traffic in an area of widely spaced apartment buildings at the edge of a large city (UTZ V), with an average height of 20 m and average spacing of 50 m (recall from Figure 4.2 that this mix of building heights/separation makes up about 29 percent of the area of the cities studied), it now has a maximum range of only 6.8 nmi. Figure 4.4 depicts how our sensor platform would perform over the urban terrain described above.

The difference in performance is dramatic. The sensor platform is looking for moving vehicles and must be able to see over buildings and into the city streets to do so. If we assume that traffic moving one way on a given city street is roughly as heavy as traffic moving the other way, then our sensor platform only has to be able to see half of the street to detect some movement. However, even this reduced-coverage requirement is very restricting. It means that our sensor platform must be able to view a spot on the ground just 25 m away from a 20-m-tall building. The *minimum* depression angle for this view is 38°,[9] an angle that intersects the ground just under 6.8 nmi from our aircraft—only 1.5 nmi beyond the minimum range of our system. For targets farther away than 6.8 nmi, the platform can no longer see half of the street, because intervening buildings block the LOS. Targets closer than 5.3 nmi cannot be seen because the maximum depression angle of the radar beam is 45°, which intersects the ground at 5.3 nmi.

---

[8]The calculations here assume a flat earth so that simple trigonometry can be used. The actual angles are slightly different at long ranges such as 100 nmi. However, even at 100 nmi, the differences are quite small; at shorter ranges, there is almost no difference at all.

[9]Depression angle is derived from simple trigonometric calculations similar to those in Appendix A.

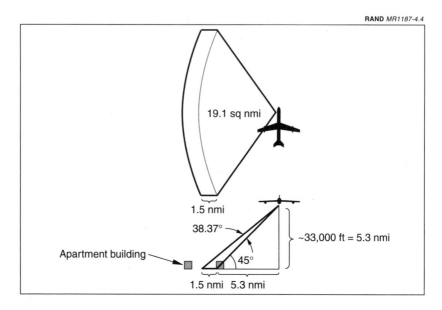

Figure 4.4—Surveillance/Reconnaissance Platform on Urban Terrain

Therefore, over urban terrain, the same sensor platform that performed so well over open, flat terrain is only able to monitor an area of about 19 square miles— less than two-tenths of 1 percent of the area it could monitor on open, flat terrain—and it must fly close to the city, potentially within range of SAMs, to do so. For comparison, even a moderately large city such as Stuttgart, Germany, covers about 28 square nmi. Furthermore, the platform can see an area 1.5 nmi deep and over 12 nmi wide. Such a thin strip does not match the typical shape of urban concentrations well and would make systematic surveillance even more difficult and time-consuming.

Even if we set aside the LOS problems just discussed and assume that somehow our sensor platform could get a clear view of all the city streets, our cueing problems are not solved. Detecting a large string of moving vehicles 100 nmi deep in enemy territory during a conventional conflict is extremely useful. Such a string of moving vehicles is almost always military. Knowing the number of vehicles, and their speed and direction, is of great value to friendly commanders: It allows them to anticipate an adversary's plans and even to attack the formations before they are able to reach the front.

In an urban peacekeeping mission, by contrast, being able to see all of the moving vehicles in a city is of less immediate value. The vast majority pose no threat to the peace U.S. forces have been sent to maintain. They simply represent the normal economic and social activity of the city. A very small number may carry guerrillas, arms, explosives, etc., but which moving vehicles those are will not be apparent from the radar display.

Urban Sensor-Cueing Concept. In short, a single large sensor platform is not adequate for sensor or target cueing in an urban environment. We need to think differently about how to perform sensor-cueing tasks in an urban setting. Instead of a single large, long-range sensor platform, the short sight lines of urban settings require a mix of small, distributed air- and *ground*-based sensors to detect and track small, distributed targets in the urban clutter. The unattended ground sensors would also perform the critical task of cueing mini-UAV and other aerial sensor platforms. These sensors should take advantage of radar, seismic, acoustic, chemical, multispectral image processing (MSI), electro-optical (EO), infrared (IR), and magnetic phenomena.[10] Since these sensors would rely on such a wide range of phenomenologies and would, in many cases, be very close to targets, they could augment UAV-based sensors if smoke, weather, or adversary countermeasures made aerial observation difficult or impossible. Figure 4.5 illustrates how such a sensor-cueing concept might work.

First, unattended ground sensors would use simple acoustic, seismic, EO/IR, laser, and radar sensors deployed in the city from the air or by friendly ground personnel. The sensors collect data and transmit it to a high-altitude, long-endurance UAV. Most of the data these sensors collect will consist of simple information, such as a count of the

---

[10]Chapter Six explains how these and other sensors could be used to support the sensor-cueing concept discussed here. For now, it is worth noting that line-of-sight limitations would require an urban sensor network made up of hundreds or thousands of UGS, with the highest density of sensors in the urban core.

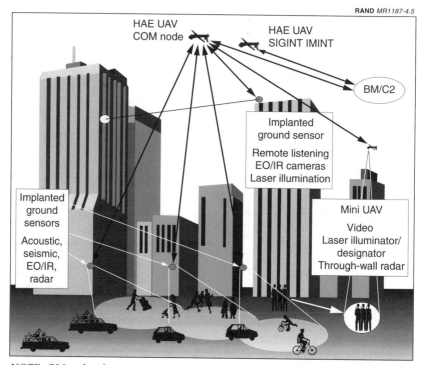

NOTE: BM = battlespace management; C2 = command and control; COM = communications; HAE = High Altitude Endurance; IMINT = image intelligence; SIGINT = signals intelligence.

Figure 4.5—Urban Sensor-Cueing Concept

number of people or vehicles that have passed a given point, or an alert message that a person with a large, concealed metal object that could be a gun just passed the sensor. However, the sensors might also be able to transmit periodic snapshots of the areas they are looking at.[11] In addition, some sensors might be capable of operating

---

[11]Snapshots from these sensors are preferable to continuous video, for three reasons: (1) They require far less bandwidth to transmit, (2) they consume less of the sensor's limited battery supply, and (3) analyzing continuous video from hundreds or even thousands of unattended sensors could require almost as many human imagery interpreters at the fusion center.

as video cameras for short periods if, for example, smoke made observation from aerial platforms difficult.

The UAV relays the sensor inputs to a battle management/command and control facility. At this facility, the data are analyzed to look for human- and vehicle-traffic patterns, classify vehicles, find people carrying weapons, etc. Data from different sensors looking at the same potential target are compared to reduce false alarms. This information is then used to cue mini-UAVs flying at low altitudes. Using high-resolution EO sensors, the operators of these mini-UAVs would positively identify and track targets.

Another important use of the data collected by these sensors is to compile a longitudinal database so that human- and vehicle-traffic patterns can be established and anomalies in current traffic investigated. Changes in traffic patterns could serve as important early-warning signals for U.S. forces, indicating that segments of the civilian population are avoiding areas they normally frequent because they have advance knowledge of ambushes, car bombs, or other violent acts. In addition, traffic-pattern analysis could serve as yet another cue for high-resolution EO sensors.

Recall that, in many peace operations, the goal is to monitor activity in open areas and detect specific threats to the normal social and economic activity of a city. A sensor network like the one described above would be a good first step toward improving the ability of aerial forces to contribute more substantially to urban missions. However, once suspicious activity has been detected, specific targets must be identified, tracked, and eventually engaged or neutralized if normal activities are to be protected. The next subsection discusses some of the challenges of identifying and tracking targets in an urban environment.

Target Identification/Tracking in the Urban Environment

The target identification and tracking task in the urban environment is very demanding. Cities are full of moving people and vehicles. Simply seeing a group of people or a vehicle moving in the city is not enough. Once some suspicious or unusual activity is detected by the sensor cueing network described above, high-resolution EO sensors take on the critical tasks of determining what exactly is happening, whether it poses a threat, and who is involved, and then track them. To do all this, the sensor operators or target-recognition software must be able to distinguish between the following:

- Civilian vehicles of similar makes, models, and colors

- A civilian construction crew repairing pipes under a roadway as opposed to guerrillas laying mines

- A plumber with a pipe as opposed to an urban guerrilla with a rifle

- An adversary soldier with an AK-47 as opposed to a friendly soldier with an M-16

- A civilian looking out a window as opposed to a guerrilla waiting in ambush.

These few examples show the kinds of subtle and demanding distinctions urban sensor operators will have to make if they are to be effective in the urban environment. They will require National Imagery Interpretability Rating Scale (NIIRS)- 9 quality imagery and intensive training in order to accomplish their demanding task. Owing to atmospheric attenuation, weather, and other factors, sensor platforms must be below about 5,000 ft above ground level (AGL) and within 2 to 3 kilometers (km) of their targets to provide images of this quality.[12]

One way to position sensors below 5,000 ft AGL and within a few thousand meters of their targets is to use UAVs. Many UAVs, such as Predator, have long endurance (up to 24 hours over a target 500 nmi from the Predator base) and keep human operators out of harm's way. Visible to the unaided eye below about 7,000 ft AGL and audible

---

[12]For a discussion of these and related issues, see Alan Vick, David T. Orletsky, John Bordeaux, and David A. Shlapak, *Enhancing Air Power's Contribution Against Light Infantry Targets*, Santa Monica, Calif.: RAND, MR-697-AF, 1996, Chapter Three. This document also contains the following description of the NIIRS scale: "A variety of metrics is used to measure the quality of an image (e.g., ground-resolved distance). The National Imagery Interpretability Rating Scale (NIIRS), developed by professional photo interpreters, is the standard used in the intelligence community. It takes into account image sharpness, contrast, and other factors, rating images on a scale from 0 to 9. For example, an image of an enemy airfield in which taxiways and runways could be distinguished would be NIIRS 1. At the other end of the scale, an image in which vehicle registration numbers (i.e., license plates) on a truck could be read would receive an NIIRS 9 rating" (p. 21, fn 16).

below 4,000 ft, Predator is reasonably large, unmaneuverable, and slow-moving, making it vulnerable to a wide range of weapons at low altitude.[13] Predator may be suitable for some low-threat situations. However, as threat levels increase, a different type of UAV will be required.

Smaller UAVs. What is needed is a smaller, quieter platform that will be invisible and inaudible at the ranges and altitudes required: a mini- or micro-UAV to serve as an urban EO sensor platform. Something between the size and capability of the two Naval Research Laboratory (NRL) mini-UAVs might work well (see Table 4.3).

The NRL Sender was developed to carry EO sensors, chemical-/ biological-agent detectors, or a tactical jammer. It can be recovered via a belly-skid landing but is cheap enough to be expended if necessary. The Swallow is an experimental UAV with a biological-agent-detector payload. Both UAVs have electric motors and a limited endurance of about 2 hours.

It should be possible to develop a UAV with dimensions between these two that could operate at approximately 1000 m (3,300 ft) AGL. Such a mini-UAV could be powered by a well-muffled piston engine

Table 4.3

UAV Dimensions

	UAV		
Specification	NRL Sender	NRL Swallow	GA Predator
Power Plant	300-W DC motor	1,500 W DC motor	85-HP 4-stroke
Wingspan (ft)	4	15	49
Length (ft)	4	6	27
Height (ft)	1	2	~7
Max Payload (lb)	2.5	10	450
Operating Speed (kt)	50	~55	~70
Operating Height (ft AGL)	1,000	10,000	~15,000

[13]See Vick et al., 1996, Chapter Four.

or a rechargeable battery and electric motor.[14] It would have a wingspan of perhaps 6 to 8 ft and a payload of about 5 lb, which would allow it to carry a high-resolution EO sensor and a laser rangefinder/designator on a stabilized gimbaled mounting. It would be sufficiently small and quiet to remain undetected by unaided human eyes and ears at its operating altitude. In addition, with an operating speed of about 50 kt at 25° of bank, these small UAVs would have very small turning radii, about 145 m, and would make one complete orbit in about 35 sec. At 1,000 m AGL and a kilometer or two from the target, these small orbits would result in very little sensor sight-line change and would allow the mini-UAVs to maintain continuous coverage of targets in streets or of particular windows, almost as if they were hovering. UAVs of the size, and with the capabilities under discussion here, could accomplish many of the tasks listed at the beginning of this section, at the altitudes and ranges just mentioned.

This approach has some difficulties. The first is choosing an operating altitude that would get the sensor as close to the target as possible to maximize image quality, which would require the UAV to fly as low as possible. However, flying lower increases the chance that the platform will be spotted from the ground and perhaps shot down. In addition, flying lower decreases the area in which the sensor platform has LOS to streets or building faces (see below). So, the trick is to find an altitude at which sensors have sufficient resolution, the platform is invisible and inaudible from the ground and high enough to cover the widest possible area of the city. These difficulties make it impossible for UAVs like those we have described here to accomplish some very demanding tasks, such as identifying individual people, at altitudes and distances where they have any reasonable chance of remaining covert.[15]

However, it is still possible that human observers might be able to detect and attack the mini-UAVs, especially if the platforms loiter over a particular target for very long. One way to respond to this

---

[14]See Appendix B for a discussion of how electric-powered mini-UAVs could be recharged by larger UAVs while in the air, using high-power microwaves.

[15]See the discussion of Figure 6.5 in Chapter Six for more on the requirements for identifying individuals from aerial platforms.

problem might be to build a large number of low-cost decoys. The decoys would have the same structure and engine as a real mini-UAV but would have no sensors, communications link, or other payload. Instead, they would simply be programmed to fly to, or orbit around, pre-selected Global Positioning System (GPS) coordinates, giving an adversary the impression that U.S. surveillance platforms were more numerous than they actually would be. This subterfuge could suppress adversary activity in the open over a much wider area than could actually be viewed at any given moment. In addition, if an adversary chose to engage the mini-UAVs with air defense weapons, many missiles would be wasted.

VTOL UAVs. One possible alternative to fixed-wing UAVs would be to use vertical takeoff and landing (VTOL) UAVs to identify and track targets in an urban environment. These would have the advantage of being able to hover and, if conditions permit, land on rooftops or other areas to observe stationary targets, important intersections, or buildings. However, it is not clear that they can be made quiet enough to remain covert as they land on or near occupied buildings. Moreover, unless they are built to look like a rooftop vent or some other part of a building, they may be easy to detect while perched.[16]

The discussions that follow assume, unless otherwise stated, that the sensor-carrying UAV is flying at 1,000 m (approximately 3,300 ft) AGL and, while remaining covert, can provide imagery of sufficient quality to accomplish many of the tasks listed at the beginning of this section.

EO Sensors Versus Targets on Rooftops and in Streets

With a comprehensive network of cueing sensors such as the one described in the previous section, mini-UAVs like those described above could quickly investigate, identify, track, and, if necessary, designate any suspicious activity on rooftops throughout a city. In

---

[16]Chapter Six presents more details on, and an operational concept for using, VTOL UAVs in the urban environment.

almost all cases, the mini-UAVs will have unobstructed LOS to al-most any roof.[17] But, what about the streets?

Recall the example of the observer and the pole from the beginning of this chapter. The ability of airborne sensor platforms to see over buildings and into streets is affected by the same set of variables as for the observer on the pole: altitude above the ground, the height of intervening buildings, and street width. Unlike the earlier example in which observer height changed with different combinations of building height and street width, for most of this section we will keep sensor platform altitude constant at 1,000 m AGL and see how different building-height and street-width combinations affect how far the sensor can see. Before examining how these real-world combinations affect what can be seen from the air, we can get a feel for what to expect by looking at an example of a fictitious city that is made up predominantly of 100-m-square blocks, all with 15-m-wide streets and over which a sensor platform is flying at 1,000 m AGL (Figure 4.6).

Since all of the streets are the same width and our sensor platform is flying at a fixed altitude over the city, the only variable left to affect how far our UAV's sensors can see is building height. The sensors can see three-fourths of a 15-m-wide street over a 1-story building, out to about 1100 m.[18] Over a 5-story building, they can see three-fourths of the street out to about 225 m. Over 10- and 20-story buildings, they can see out to about 113 and 57 m, respectively.[19, 20]

---

[17]A variety of existing USAF sensor platforms, such as the AC-130 and Predator, could accomplish the rooftop surveillance mission. However, as previously mentioned, these platforms would probably not have sufficient sensor resolution to identify many potential targets and would therefore need something like a mini-UAV or VTOL UAV to provide imagery for target identification.

[18]The ability to view three-fourths of the space between buildings is probably suffi-cient for most surveillance and reconnaissance tasks. While some important activity could occur on the fourth of the street the UAV cannot see, virtually all vehicles, and many people, attempting to move through the "unseen" area would be at least partly visible to the UAV. In many cases, the top half or more of people or vehicles will be visible when their bottom halves are blocked from view by buildings. Such a view, es-pecially of people, will often be sufficient to allow for identification and tracking.

[19]See Appendix A for a short explanation and example of the simple trigonometry by which these numbers were calculated.

[20]It is important to note that, for the purposes of illustration, this figure assumes buildings of a uniform height. In any real city, the possible line of sight would be much

RAND *MR1187-4.6*

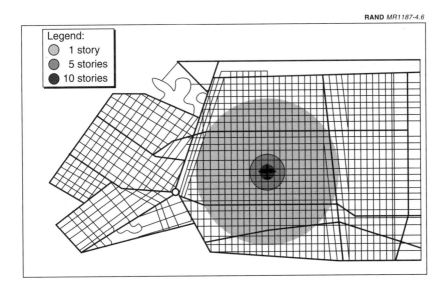

Figure 4.6—Airborne Sensor View of Three-Fourths of 15-m-Wide Streets over Buildings of Different Heights, from 1,000 m AGL

With the preceding discussion as a foundation, we can now move on to examine how the low-flying UAV sensor platforms described above would perform against targets in a city of typical size with a typical mix of UTZs. One way to do this is to take the mean UTZ areas for the 14 cities studied by Ellefsen (1999) and combine them into a notional city we will call Averageburg.

The total area of Averageburg is 5,514 hectares, or 55.14 square km.[21] Table 4.4 shows how these 5,514 hectares are divided among the seven UTZs. It also shows what percentage of total Averageburg area each UTZ accounts for, the average building height and street width within each UTZ, and how many surveillance UAVs it would

---

more jagged because of variation of building heights both within, and especially between, UTZs.

[21]A *hectare* is a metric measure of land area, 100 × 100 m. There are 100 hectares in a square kilometer, so 5,514 hectares is 55.14 square km.

Table 4.4

**Average UTZ Areas, Attributes, and UAVs Required for Line of Sight
to Any Part of Each UTZ**

	UTZ I	UTZ II	UTZ III	UTZ IV	UTZ V	UTZ VI	UTZ VII
Area (hectares)	140	61	527	528	1641	1798	819
Percent	3	1	10	10	29	32	15
Avg. Height (m)	30	45	6	9	20	6	9
Avg. Street Width (m)	5	30	20	25	50	15	70
No. UAVs Req.	252	8	3	5	14	18	1

SOURCE: UTZ data are from Richard Ellefsen, *Current Assessment of Building Construction Types in Worldwide Example Cities*, Report prepared for Naval Surface Warfare Center, Dahlgren Laboratory, Dahlgren, Va., 1999, Appendix 12.

take to have line of sight to all streets (using the same three-fourths view minimum as above) in each UTZ. The final row of the table shows how many UAVs are required for each UTZ, so that if a coordinate were chosen at random in the city, one of the UAVs would have line of sight to it.

However, these UAV numbers are based on some unrealistic assumptions—for example, that all of the buildings and streets in each UTZ are the average height and width, and that each zone is made up of a number of areas that perfectly match the surveillance capability of our UAV for each zone. But to assume that UTZ V is made up of 14 circular areas exactly 625 m in diameter so that only 14 surveillance UAVs are required to have LOS to rooftops or at least three-fourths of any street in the zone has little relationship to any real city.

The actual number could be either more or less than those shown. It could be more, because in any real city the shape and size of UTZs will not conform to sensor-platform capabilities as assumed here. Inefficiencies will result, because some areas are double-covered to eliminate gaps. And it is unlikely that continuous LOS to all streets and rooftops in a city would be needed. A short delay and some gaps in coverage would enable UAVs to be moved to where they are needed at up to 100 kt (their maximum speed). We might be able to get by with far fewer UAVs covering Averageburg. No urban air-and-ground-based sensor network such as the one described earlier exists, and no one knows how different mission requirements, target types, target exposure times, smoke/haze, and other factors would alter the number and type of UAVs and unattended ground sensors required to monitor a given urban area. Obtaining more accurate estimates of the number of UAVs and/or unattended ground sensors required to monitor different UTZs will require extensive modeling/simulation efforts allied with experimentation with prototype sensor systems to provide empirical inputs for the modeling and simulation effort.

The purpose of these numbers is not to accurately predict how many surveillance UAVs would be required to monitor a city but to show that, owing to the extremely steep viewing angles required—81° is the minimum average viewing angle in UTZ I—aerial platforms must be almost on top of targets in UTZs I and II. Therefore, the bulk of the air and unattended ground sensor effort expended in the average city will be focused on UTZs I and II. These areas, the city core, account for only 4 percent of the land area of Averageburg but 260 of the total of 301 UAVs (86 percent) required to watch the rooftops and streets of Averageburg. The tall, closely spaced buildings in UTZ I alone require 252 UAVs, or almost 84 percent of the total. This 81° viewing-angle requirement results in very short maximum sight lines for UAVs operating over this UTZ. UTZ II requires minimum viewing angles of about 57°. In contrast, the remaining 96 percent of Averageburg requires viewing angles of about 20° or less and only about 40 UAVs, or about 14 percent of the total.[22]

Outside the city core, it may be possible to maintain EO sensor coverage of an average city with only a fraction of the resources devoted to monitoring the city core. Inside city-core areas, sight lines are so short that aerial surveillance and reconnaissance of lower floors of buildings and ground targets is extremely difficult and resource-intensive. For high-resolution surveillance and reconnaissance of high-priority targets in these areas, it will probably be necessary to

---

[22]Minimum viewing angles were computed using the street-width and building-height data from Table 4.4 and trigonometric calculations like those shown in Appendix A.

supplement aerial platforms with unattended ground sensors on roofs and sides of tall buildings.[23]

Note that, while the UAVs can see almost anything of interest in the streets or on rooftops in the areas they cover, they cannot see everything at once. In fact, if they are to deliver high-quality NIIRS-9 imagery, they will have to use powerful zoom or telephoto lenses with very narrow fields of view. This explains why the sensor cueing network described in the previous section is so critical for effective urban surveillance and reconnaissance.

EO Sensors Versus Targets Inside Buildings

The geometry associated with attempts to view targets inside buildings from airborne sensor platforms is similar to that associated with looking at targets in the streets or on rooftops, but more complex.

In some ways the demands are less stringent. Figure 4.7 shows how, for buildings of a given height on a street of a given width, looking at targets in buildings allows lower depression angles, and much greater UAV horizontal standoff distance, than looking at targets in streets.

The figure shows two UAVs looking over the same building. The UAV at the top is looking for targets in the street; the lower UAV is looking for targets in windows of the building on the left. The UAV looking for targets in windows can fly lower and/or farther away, because it is able to take advantage of the full street width, which allows much lower minimum viewing angles—about 22° as opposed to about 60° for UTZ V, for example—when looking at building faces rather than in the street.[24]

---

[23]See Chapter Six for a description of how these sensors might be delivered covertly and used.

[24]These angles were derived from trigonometric calculations similar to those presented in Appendix A.

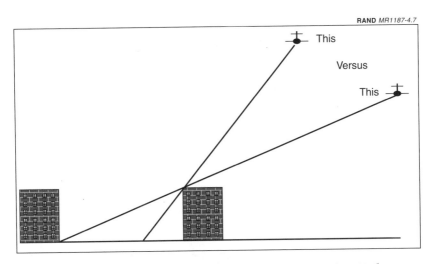

**Figure 4.7—Targets in Buildings Allow Lower Depression Angles
Than Do Targets in Streets**

However, a number of complications are associated with looking for targets inside buildings with EO sensors that do not apply to looking for targets in the street: openings to look through, shifts in interior views as vertical viewing angle increases, building side closest to platform.

Openings. There must be an opening in the building to look through. This will most commonly be a window, and the discussion that follows assumes that it is, although it could also be an open door or any other opening in a wall. This can be a real problem in most parts of a typical city.[25] Despite their shortness and relatively wide streets, which allow low viewing angles, the buildings typical of UTZs III, IV, VI, and VII have less than 25 percent of their wall area devoted to venting, or windows and doors. Together, these UTZs account for about 45 percent of a typical city's wall area (Table 4.5).

---

[25]In the long run, other sensor technologies, such as through-the-wall radar, might be used to supplement, or in some cases replace, EO sensors for targets inside buildings. Chapter Six describes these technologies.

Table 4.5

Building Attributes by Urban Terrain Zone

UTZ Type	Percent of Typical City Area (%)	Percent of Wall Area (%)[a]	Avg. Height (m)	Avg. Separation (m)	Approx. Venting (%)
I	3	7	30	5	70
II	1	2	45	30	90
III	10	14	6	20	15
IV	10	8	9	25	5
V	29	45	20	50	70
VI	32	16	6	15	25
VII	15	7	9	70	5

SOURCE: UTZ data are from Richard Ellefsen, *Current Assessment of Building Construction Types in Worldwide Example Cities*, Report prepared for Naval Surface Warfare Center, Dahlgren Laboratory, Dahlgren, Va., 1999, p. 28 and Appendix p. 12.

NOTE: The percent of city area column states how much of the land area of the city is accounted for by each UTZ. The percent of wall area column differs from the percent of city area because some UTZs have greater average building heights, and therefore more wall area, than do others. In general, UTZs with greater average building height account for more of a city's wall area than its land area. The only exception is UTZ III, attached houses, which has a greater wall percentage than area percentage because of the extreme density of the attached buildings in this UTZ.

[a]Because of rounding, this column does not add to 100 percent.

UTZs IV and VII (closely and widely spaced industrial storage buildings, respectively, with only 5 percent of their wall area devoted to windows and doors, offer particularly poor prospects for interior viewing. UTZ III (attached houses near the city core) is not much better. Only 15 percent of the wall area for these buildings is devoted to windows and doors. However, Ellefsen notes that the 15-percent figure is for *front and rear walls only,* so the actual figure is probably closer to 7–10 percent.[26]

UTZs I (closely spaced inner-city buildings), II (widely spaced highrise office buildings), and V (widely spaced apartment buildings) are made up mostly of modern frame-construction buildings with extensive windows, from 70 to 90 percent of the wall area. However, as the

---

[26]This makes sense, because these buildings are attached to their neighbors on either side.

next part of this discussion illustrates, particular problems are associated with the high depression angles required to see through the lower windows of these tall modern buildings, especially those in UTZs I and II at the city core.

Both horizontal and vertical viewing angles greatly affect the perceived size of windows and other openings in buildings, a factor that is critical for two reasons. First, windows are apertures into rooms, so the size they are perceived to be directly affects how much of the room, theoretically, can be seen from an aerial platform. Second, aerial weapons often will have to enter rooms through existing openings, so a smaller perceived window (opening) size will make weapons delivery more difficult.[27]

Figure 4.8 illustrates how perceived window size changes as a function of vertical viewing angle. When the observer is level with the

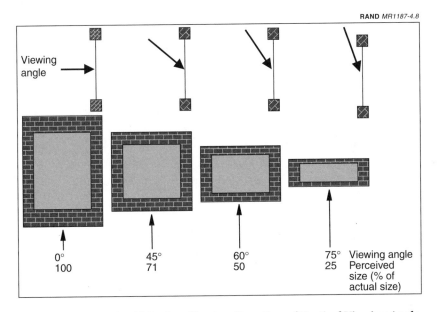

Figure 4.8—Perceived Window Size As a Function of Vertical Viewing Angle

---

[27]Weapon-delivery issues are discussed in more detail in the next section.

window (viewing angle of 0°) the window *seems to be* its full height. As the observer's altitude above the window increases, so does the viewing angle. At 45°, the window appears to be only 71 percent as tall as it did at 0°. At 60° vertical viewing angle, the window appears to be only half as high as at 0°. At 75°, it seems to be only 25 percent as tall. As viewing angle increases further, perceived window size rapidly decreases, so that it is zero at 90°.[28]

The average building heights and street widths in UTZ I require vertical viewing angles of about 80° to see into lower floors from airborne platforms. UTZ II requires viewing angles of about 56° to see lower floors.[29] In contrast, most of the other UTZs, even UTZ V, allow much lower minimum viewing angles of about 17° to 22°. The exception is the widely spaced short storage buildings of UTZ VII, which allow viewing angles as low as about 7°.

Shifts in Interior Views. Small perceived window size is not the only problem associated with high vertical viewing angles. Interior views shift as vertical viewing angle increases. Imagine a sensor platform level with a window and directly in front of it. The sensors can see all the way to the back of the room. As the platform altitude increases, the terminus of the sensor's view shifts down the wall and onto the floor. As viewing angle continues to increase, the terminus shrinks with perceived window size and moves closer and closer to the window. At 75°, or even 60°, the sensor is sweeping only a very small area of the room within a few feet of the window. Figure 4.9 illustrates this process.

The same perceived window size and interior view phenomena are at work in the horizontal dimension, making matters even worse. Figure 4.10 shows how these phenomena interact. Choosing a particular elevation angle, say, 45°, allows the perceived window size at various azimuths to be determined. For example, at 45° elevation and 45° azimuth, the window seems to be only half its actual size.

---

[28]Angles are measured from the observing UAV's perspective, with angles increasing from the horizontal (0°) to full vertical (90°).

[29]Airborne sensors will often enjoy much more modest viewing angles to the upper floors of tall buildings, which could make them much more effective against targets, such as snipers, on the upper floors of tall buildings.

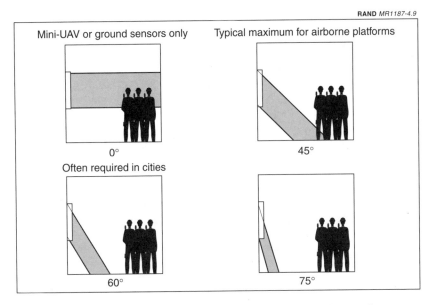

Figure 4.9—Interior Views Shift with Vertical Viewing Angle

One way to improve the chances UAV sensor platforms see something useful when they point their lenses at windows is to position the platforms so that they maximize their perceived window size. This could be done by avoiding azimuth and elevation combinations that result in perceived window sizes below, for example, 50 percent. With a few exceptions at low azimuths, this practice would essentially limit sensor platforms to the upper-left-hand quadrant of the various combinations of azimuth and elevation shown in Figure 4.10, or to combinations where *both azimuth and elevation are less than 45˚.*

How do we achieve this positioning? We can maintain a particular elevation angle by flying a constant altitude, recognizing that much of the urban core cannot be seen at that angle. Horizontal angle is trickier, since it changes as the surveillance platform approaches, then passes, the target. Unless, the surveillance platform flies a

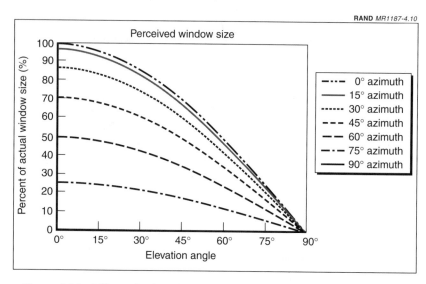

Figure 4.10—Effect of Azimuth and Elevation on Perceived Window Size

racetrack pattern (using a gimbaled sensor to maintain coverage), optimal viewing angles (–45° to +45°) may be quite fleeting as the UAV zips by the target. How fleeting depends on standoff distance and speed. Unfortunately, high-resolution surveillance forces our platform closer, where angle relative to the target changes more quickly. A mini-UAV flying a racetrack pattern or a VTOL UAV, hovering or landed, may be the only feasible means of providing high-resolution images of an interior space from airborne platforms.

Figure 4.11 illustrates some of the implications of these azimuth and elevation limits for the area in which a UAV sensor platform would have useful LOS into windows, as opposed to in streets, as shown in Figure 4.6. Compared with Figure 4.6, the area in Figure 4.11 where the sensor can see over 1-story buildings is huge. It extends to a radius of about 4,500 m, well beyond the edge of the figure at this scale and 4 times as far as the same UAV at 1,000 m AGL could see three-fourths of the street over a 1-story building.

However, this is where the direct comparison between the two figures ends. Unlike Figure 4.6, Figure 4.11 shows how far the UAV could see over 2-, 3-, and 4-story buildings when looking at building faces. This is the limit of its capability. In order to look over buildings of five stories or more in height on 15-m-wide streets, the UAV would have to look down at an angle greater than 45°.

From the preceding discussion, we know that angles greater than 45° present increasing difficulties for surveillance of targets inside buildings; therefore, Figure 4.11 assumes there is a hole in sensor coverage of building faces when the angle to the ground is greater than 45°. This is not as fatal a limitation as it might seem. Ninety-six percent of the city area has building-height-and-street-width combinations that allow LOS at depression angles of less than 45°. In addition, within this 96 percent of city area, the fact that sensor platforms looking into windows can take advantage of the full width of the street allows them to see their targets over much greater areas than sensors at the same altitude looking for targets *in* the street.

No View of Building Side Closest to Platform. Another constraint on aerial sensor platforms attempting to look into buildings is that they cannot see any of the windows in the building faces on the side of the

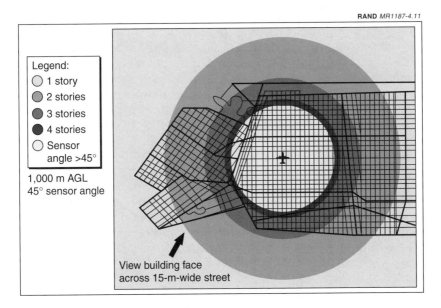

Figure 4.11—View of Building Faces Across 15-m-Wide Streets

street closest to the sensor platform. If the platform is looking north, it can see the south-facing windows on the north side of a street only, not the north-facing windows on the south side of a street. The same applies to other directions as well.

Summary and Conclusion. In theory, for most UTZs, a sensor at 1,000 m looking at building faces can see an area at least 10 times larger than the same sensor can see when looking at streets. So, even with the doughnut-shaped sensor coverage area depicted in Figure 4.11 and the other limitations discussed above in mind, the number of UAV sensor platforms required to view targets in the street will probably drive sensor-platform numbers for most cities.[30] This is just as well, since aerial sensor platforms are likely to be much more effective against targets in the streets than against targets in windows, even with excellent cueing and all of the considerations just discussed.

Many missions will take place in intact cities. Therefore, even if the strict geometric limitations discussed above are observed, sensors will still have to contend with reflection off windows, tinted windows, shadows within rooms, and curtains and blinds that will hide much of the activity taking place behind windows within LOS. In addition, buildings in about 55 percent of a typical city's wall area will either require extreme depression angles or have 25 percent or less window area in their outer walls. Therefore, there is little promise that airborne platforms using electro-optical sensors can effectively observe building interiors in most built-up areas.

In most cases, alternative sensor platforms or other means of surveillance will be required. Unattended ground sensors, delivered either from the air or by ground personnel, are one option for getting sensors close enough to detect, identify, and designate targets inside a building. Another option is to use teams of ground personnel. Still another is to use VTOL UAVs that land on nearby buildings, observe, and then move on. Finally, mini- or micro-UAVs small enough to fly down streets at window level may be an option for some missions. These options are discussed in detail in Chapter Five.

---

[30]Just as with Figure 4.6, Figure 4.11 assumes buildings of a uniform height for the purposes of illustration. In any real city, the possible line of sight would be much more jagged from variation of building heights both within and, especially, between UTZs.

This section has explored some of the issues associated with aerial surveillance and reconnaissance in urban environments. It has examined in detail the sensor cueing problem, as well as the challenges of identifying and tracking targets in streets and inside buildings. These are important tasks in and of themselves. They are also necessary first steps prior to delivering weapons in those many cases when potential targets prove to be hostile. The next section describes the challenges urban terrain presents to conventional aerial weapon delivery.

AERIAL WEAPON DELIVERY IN THE URBAN ENVIRONMENT

U.S. aerospace power has greatly improved its ability to strike specific urban targets in strategic air campaigns directed at targets deep in an adversary's homeland. This is an important capability that continues to improve. However, many of the weapons that are so effective at destroying large buildings, bunkers, bridges, etc., with minimal collateral damage in such campaigns, are not suitable for attacking targets often encountered in more limited urban operations, for two primary reasons: incompatibilities between existing weapon trajectories and delivery angles dictated by the geometry of urban terrain, and the explosive yield of the vast majority of current USAF weapons, which would produce unacceptable risks of collateral damage and/or fratricide if used in many urban situations.[31]

Aerial Weapon Trajectories and the Geometry of Urban Terrain

The angle at which a weapon strikes a building wall or roof is of great importance. The best angle for a weapon to strike a flat surface such as a wall or roof is 90° (perpendicular to the surface). At this angle, the kinetic energy of the weapon is concentrated on the smallest

---

[31]Notable exceptions to this general rule are the 105mm, 40mm, and 25mm guns carried by AC-130s. Individual rounds from these weapons are accurate and (with the possible exception of the 105mm) generally have room-sized rather than the building-sized effects typical of other aerial weapons. This makes AC-130s far more suitable for many urban targets than other USAF weapon-delivery platforms. However, the proliferation of advanced MANPADS and the vulnerability of the AC-130 to these systems may make it difficult for AC-130s to operate at altitudes where they can be both safe and effective.

possible area, maximizing the possibility of penetrating the wall or roof. In addition, the velocity vector of the weapon at 90° to the target surface ensures that there is negligible chance that the weapon will glance off. It may fail to penetrate if the wall or roof is strong enough, but it will not be deflected.

Impact angles of 90° are almost impossible to achieve with air-delivered bombs against walls. While some laser-guided bombs (LGBs) can maintain level flight, or even climb, for short periods, the terminal phases of their flights are governed by the forces that affect any bomb: gravity, forward velocity at the time it was dropped, and aerodynamic drag. In general, these forces combine to cause LGBs dropped from level flight by modern U.S. fighters to fall at an angle of between 15° and 20° (relative to the ground); therefore, they impact at angles of about 70° to 75° relative to the wall. This is not really a problem, since the probability a projectile will glance off a wall does not markedly increase as long as impact angle relative to the wall is above about 60° (30° or less relative to the ground).

Attacking horizontal targets, such as roofs, at exactly 90° with LGBs is also very difficult, but for different reasons. Many LGB guidance systems contain small auto-pilots that use inputs from gyroscopes and the laser seeker to control movement of the bomb's canard control fins. If the bomb's attitude is too close to vertical the gyroscopes tend to tumble, sending erroneous inputs to the auto-pilot, which, in turn, sends erroneous inputs to the control fins, sending the bomb out of control. To avoid this and still allow the bomb some freedom to make downward corrections, the maximum dive angle for most LGBs is about 60°.[32] For the same reasons as noted above, this angle is perfectly acceptable for attacking most horizontal targets.

Figure 4.12 illustrates the standard LGB attack angles for vertical (wall) and horizontal (roof) targets. It also shows the 17° vertical target attack skimming over the roof of a smaller building to hit the lower floor of the multistory apartment building on the left, which is appropriate. Figure 4.13 shows that even these shallow deliveries can be used against more than a third of the walls in a typical city, and against more than half in some cities such as San Jose, Costa Rica.

---

[32]Some newer guided munitions, such as the Joint Direct-Attack Munition (JDAM) have the ability to attack targets at much steeper angles.

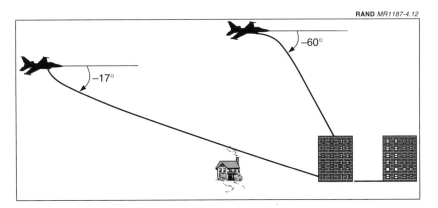

RAND *MR1187-4.12*

Figure 4.12—Standard LGB Deliveries for Vertical and Horizontal Targets

Figure 4.13 is based on the Ellefsen (1999) building height and separation data presented earlier in this chapter. These data, combined with simple trigonometry, reveal that weapon-delivery angles as shallow as 17° are steep enough to attack many areas in a typical city. However, the data also show that, in some cities, such as Tel Aviv, the standard 17° LGB vertical target bomb trajectory is not steep enough to hit the vast majority of building walls in the city.

The figure systematically underestimates the ability of 17° bomb deliveries to hit city walls. To maximize the delivery angle required to hit a target for each combination of building height and street width, the estimates assume airborne attackers are always shooting at targets on the lowest floors of buildings. If delivery angle is greater than 17°, the figure assumes a 17° delivery cannot hit any of the walls in a given UTZ. Deliveries of 17° could probably be used successfully against targets on the upper floors of buildings in many situations where the figure assumes they cannot. The same bias exists for 30° and 60° delivery calculations, but it is smaller. At these angles, even with this conservative estimation method, over 90 percent of the wall area of a typical city is vulnerable to attack.

Figure 4.13 also shows there is little to be gained by increasing delivery angles above 30°. Attacking at angles as high as 60° does little to increase the proportion of wall area that can be attacked, but greatly

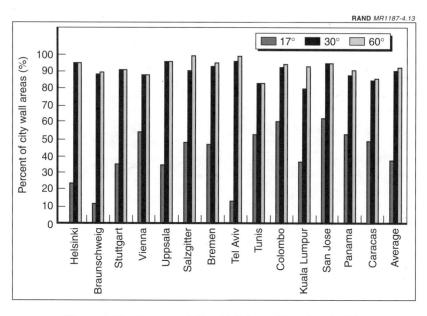

Figure 4.13—Percent of City Wall Areas That Can Be Hit

increases the chances that a weapon will be deflected off the structure's walls and land in the street. In addition, even if a weapon penetrates the wall on a 60° trajectory, the weapon's steep angle means that it will be much more likely to penetrate the floor of the room it was aimed at, increasing the chances of killing innocent people or friendly troops below. This brings us to the issue of weapon effects, which are discussed in the next section.

However, before moving on to that discussion, it is worth noting that the areas that cannot be attacked with 30° or even 60° weapon-delivery angles are the same city-core areas (UTZs I and II) in which aerial surveillance and reconnaissance are so difficult. As with surveillance and reconnaissance, these relatively small, but important, areas of cities demand new and different approaches to weapon delivery. Several potential operational concepts for delivering weapons deep into urban canyons are described in Chapter Five.

Weapon Effects and Urban Terrain

Weapon-delivery geometry is the first problem in delivering weapons against urban targets. The second problem is that the vast majority of weapons in the USAF inventory were designed to destroy entire buildings or armored vehicles, not produce limited effects in a single room. Their kinetic energy, blast, fragmentation, and other terminal effects are far in excess of what is required or advisable for many urban targets. Table 4.6 lists most standard USAF and many Army weapons and their warhead weights, along with an estimate of the terminal effects they produce.

The table makes it clear that the vast majority of USAF weapons are unsuitable when the situation requires that lethal effects be limited to a room or other small area. For example, the smallest standard LGB available to USAF fighter crews is the GBU-12, which uses a standard Mk-82 500-lb bomb as its warhead. If guided by a laser designator on a surveillance UAV or deployed with a ground team, this weapon can be very accurate.[33] As we have seen, it can theoretically hit many, even most, urban targets. However, since it arrives at just under the speed of sound and is packed with hundreds of pounds of high explosive, it is likely to heavily damage, or completely destroy, the vast majority of buildings in a typical city. In addition, the blast, fragments, flying glass, etc. created by the explosion are likely to damage neighboring buildings and to kill or injure people in nearby buildings and on surrounding streets. In short, this is not the type of weapon to use against a sniper on the upper floors of an apartment building, or in support of a ground patrol ambushed by attackers in surrounding buildings.

Only three currently available USAF weapons fit into the room-sized-effects category. One of them, the Low-Cost Autonomous Attack

---

[33]The ability of fighter crews to self-designate targets in a dynamic urban environment would be limited by short sight lines, the speed and turn radii of their aircraft, and onboard sensors with insufficient resolution for positive target identification. Therefore, offboard designators with the ability to observe a scene for extended periods from fairly close range, such as mini-UAVs or ground observers, are a far better choice as designators in those urban situations for which contextual information and high-resolution sensors may often be the keys to avoiding civilian casualties or fratricide.

Table 4.6
Current and New-Concept Warhead Weights and Effects

System	Warhead Weight (lb)	Service	Size of Effects
BLU-109	~2000	USAF	Large/Multiple Buildings
Mk-84	~2000	USAF	Large/Multiple Buildings
CALCM	~2000	USAF	Large/Multiple Buildings
JDAM	~1000	USAF	Large Building
Mk-83	~1000	USAF	Large Building
M117	~750	USAF	Large Building
Mk-82	~500	USAF	Building
Maverick AGM-65G	~300	USAF	Building
Mk-81	~250	USAF	Building
ATACMS Penetrator	~250	Army	Building
155mm Artillery Round	~100	Army	Small Building
120mm Guided Mortar	~35	Army	Large Room/ Small Building
105mm Artillery Round (on AC-130)	~35	USAF	Large Room/ Small Building
Hellfire Anti-Armor Missile	~20	Army	Large Room/ Small Building
Laser-Guided Training Round	~15	USAF	Large Room/ Small Building
TOW	~10	Army	Room
LOCAAS	~10	USAF	Room
30mm (A-10)	~2	USAF	Room
40mm (AC-130)	~2	USAF	Room
Guided Mini-Glide Bomb	~2	New Concept	Room
Side-Firing Mini-UAV	~2	New Concept	Room
Precision Anti-Personnel Munition	~2	New Concept	Room
Laser-Guided Grenade	~1	New Concept	Room
Hand Grenade	~1	Army	Room
Sniper Rifle	Few Ounces	Army	Individual Person
Nonlethals	Variable	Joint	Individual to Building

NOTE: AGM = air-to-ground missile; ATACMS = Army Tactical Missile System; CALCM = Conventional Air-Launched Cruise Missile; JDAM = Joint Direct-Attack Munition; LOCAAS = Low-Cost Autonomous Attack System; TOW = tube-launched, optically tracked, wire-guided [missile].

System (LOCAAS), is not yet in production. Another, 40mm gun rounds from AC-130s, may not be available for many scenarios because of survivability concerns. The third, 30mm gun rounds from an A-10, may be difficult to deliver at steep angles and in small enough numbers to limit damage to a single room. In addition, the A-10 shares many of the survivability problems of the AC-130 when operating at optimal altitudes and airspeeds over urban terrain. Figure 4.14 illustrates some of the problems associated with the terminal effects of current USAF weapons, ranging from weapons penetrating through floors and detonating on floors occupied by "friendlies" or noncombatants to the complete collapse of the building.

The USAF does have a few guided weapons with much smaller warheads than the GBU-12. For example, the USAF has a laser-guided training round. As the name suggests, the chief purpose of this device is to help teach fighter crews how to drop LGBs without expending expensive live munitions. The training round consists of a metal tube about 4 in. in diameter and 6 ft long, with a laser seeker, guidance

RAND *MR1187-4.14*

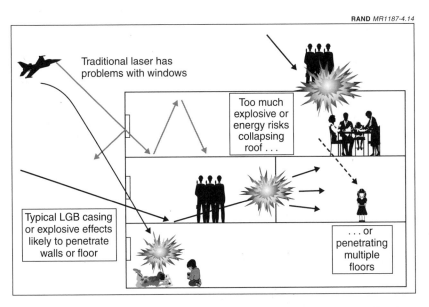

Figure 4.14—Large USAF Warheads Make Damage Limitation Difficult
for Many Urban Targets

and control unit, control fins attached to the front, and fins at the rear. It is possible to attach a small warhead to the back of the training round to give some explosive punch. The impact of this weapon is far less than that of a GBU-12; however, because it is designed to mimic its larger brethren, it arrives at the target moving near the speed of sound. An object weighing the better part of 100 lb moving at this speed has more than enough kinetic energy to smash through floors and walls or the rooms at which it is aimed.

Figure 4.14 also illustrates problems that existing laser target-designation systems have with windows. If the glass is in place, it is likely to reflect a great deal of the laser energy away from the designating platform and toward nontargeted areas such as streets, especially at steeper designation angles such as those required to attack the lower floors of city-core buildings. If the glass is not in place, the laser energy will simply pass through the window and little, if any, will be reflected back out the window at angles useful for incoming weapon guidance. (Chapter Five explores some possible solutions to this problem.)

AIRLIFT IN URBAN ENVIRONMENTS

Although this report focuses on sensor and shooter challenges in the urban setting, we do want to acknowledge the importance of airlift in the urban environment and the difficulties airlifters can face.

Fixed-wing airlifters, such as C-17s, C-5s, or C-130s, face a number of challenges. The airfield itself may be damaged by the fighting, limiting the availability of runway lighting, navigation aids, and support equipment. As the USAF experienced in Sarajevo and Mogadishu, mortars, artillery, heavy machine guns, and snipers all pose a threat to aircraft, support personnel, and passengers when the aircraft are on the ground. On approach and takeoff, small arms, machine guns, anti-aircraft artillery (AAA), and manportable surface-to-air missiles in the surrounding area threaten aircraft. There are few good counters to these threats. Friendly ground forces can help by controlling some of the approach and takeoff terrain and the more-prominent structures around an airfield, and IR countermeasures on the aircraft help some. However, at times, urban airfield operations may simply have to be shut down in the face of the more severe threats.

Airdrop operations can take the place of some airfield deliveries, but the lack of large open areas, air defense threats, and the presence of civilians or adversary military personnel will hamper them. The lack of large areas means that airdrops must be spread out over multiple passes or that some loads risk landing in woods, on buildings, or elsewhere in the city. The presence of civilians means that innocent people might be hurt by errant loads and, along with adversary personnel, pose the problem of the airdropped material falling into the wrong hands. GPS-guided parafoils or higher-speed delivery concepts (see Chapter Five) have the potential to increase the accuracy of airdrops in urban settings, avoiding these problems.

Rotary-wing aircraft are key to U.S. Army and Marine doctrine for urban operations and are relied on by all services to move personnel and resupply. Their primary problem is survivability, since they fly low and fairly slowly, making them vulnerable to small arms, AAA, and MANPADS. Although IR countermeasures provide some protection, new concepts will be necessary if rotary-wing aircraft are to survive on the urban battlefield. Optical dazzling devices, smoke, and acoustic nonlethal weapons are a few of the possible protective options that might disable opponents long enough for rotary-wing aircraft to land, take off, and leave the area. Other challenges faced by rotary-wing aircraft include city lights washing out night-vision-goggle visibility; dust kicked up during hover, producing zero visibility; obstructions to low-level flight (powerlines, towers, buildings); and debris kicked up by rotor wash. Three-dimensional maps and GPS pseudolites, discussed in Chapters Five and Six, have the potential to allow rotary-wing pilots to safely navigate between buildings when dust limits visibility.

These and related issues are complex and important and deserve detailed treatment in future studies.

CONCLUSIONS

The three main sections of this chapter presented information on urban form and how physical characteristics of the urban environment can be used to divide cities into militarily useful UTZs. In addition, the chapter analyzed how the physical characteristics of the various UTZs enhance or negate the ability of aerial platforms to conduct surveillance and reconnaissance missions and to deliver weapons. The information presented in this chapter leads to the following conclusions:

- Over 90 percent of urban land area is covered with buildings less than 20 m (about 6 stories) in height.

- Short sight lines and resulting increased depression angles relative to operations on flat, open terrain greatly restrict the capability of long-range sensor platforms to provide adequate targeting information on urban terrain.

  —A new surveillance and reconnaissance concept emphasizing unattended ground sensors for cueing and mini- or micro-UAVs carrying EO sensors for identification and tracking is required in the urban environment.

  —Developing the technology, tactics, techniques, and procedures to effectively monitor urban areas with a combination of air and ground sensors will require extensive experimentation, simulation, and modeling efforts.

- There is little promise that traditional airborne platforms can effectively observe building interiors in most built-up areas.

- Outside of the city core, standard weapon-delivery trajectories can be used to attack targets in streets and almost all buildings.

- In the geographically small but important city core:

  —Sight lines are so short that aerial surveillance and reconnaissance of lower floors of buildings and ground targets is extremely difficult and resource-intensive.

  —Extreme delivery angles prohibit use of most current weapons to attack targets where room-sized effects are desired.

  —New concepts are required to track and attack targets in the city core.

- Current USAF weapons are too powerful for many situations, even when delivery geometry would allow attacks.

  —New limited-effects munitions, delivery modes, and target-designation techniques are required to limit the probability of collateral damage and/or fratricide.

Having discussed the challenge of collecting information and delivering weapons in the urban environment, we now explore various concepts that have the potential to solve these problems and greatly enhance the effectiveness of aerospace forces in urban operations.

Chapter Five

NEW CONCEPTS FOR ACCOMPLISHING
KEY TASKS IN URBAN OPERATIONS

INTRODUCTION

As noted earlier, it is our judgment that most urban operations will fall at the lower end of the spectrum of conflict. Although there are improvements that the USAF can make in its ability to conduct urban operations against conventional foes, we see the major shortcomings as being in the ability to detect and attack unconventional foes.

It may be possible in future limited operations to identify and attack critical adversary centers of gravity or key nodes. However, historical experience suggests that this possibility will be the exception, not the rule. Rather, in most limited operations, strategic objectives are more likely to be achieved through the cumulative effect of persistent surveillance and strike than through the destruction of a small target set.

For example, in a notional peace operation that included an urban component, U.S. objectives might be to stop the violence, resettle the population, and achieve a return to normal in which routine civil and economic activities could take place without disruption. To accomplish these objectives would require, above all else, that friendly forces control the streets. The operational task of controlling the streets would, in turn, require that a variety of tactical tasks be accomplished. Among the more prominent tasks[1] are the following:

- Stop movement of combatants, vehicles, equipment.

- Provide rapid, high-resolution imagery for target ID.

- Detect and neutralize adversary ambush positions.

- Detect and neutralize snipers.

- Monitor high-priority targets.

- Resupply isolated friendly ground forces.

- Provide close support to friendly ground forces.

The most robust solution to any of these tasks involves the combination of air and ground elements working in harmony. As we look to the future, unmanned ground-sensor networks may be able to reduce the number of ground personnel put at risk, just as unmanned air vehicles are doing for air operations. Nevertheless, it will likely always be the case that it is better to have some personnel on the ground to supplement unmanned ground sensors. This chapter presents new concepts of operation to accomplish the above tasks. For each task, we discuss the nature of the challenge, then suggest a new concept of operation to accomplish it.

STOP MOVEMENT OF COMBATANTS, VEHICLES, AND EQUIPMENT

Peace operations often require that friendly forces control the city's streets. The amount of control may vary by situation; at the least, militias, irregulars, and various combatant forces must be denied the

---

[1]We also believe that the detection and neutralization of adversary manportable surface-to-air missiles will become increasingly important as more-advanced versions of these proliferate. These weapons could seriously impede all urban air operations, both rotary and fixed wing. A related task is the insertion and extraction of personnel, equipment, and supplies in the urban environment. MANPADS, AAA, heavy machine guns, and small arms are all capable of downing rotary-wing aircraft. It would be valuable to have counters to these common weapons or alternative means to move personnel, equipment, and supplies. As the October 1993 shootdown of U.S. Army helicopters in Mogadishu, Somalia, showed, even simple weapons (in this case, RPGs) can down aircraft that are flying low and slow.

freedom to mass, move, and operate, whether on foot, in modified civilian vehicles, or in armored vehicles.

For operations in which the control of the city is contested, it is vital that adversary forces not be allowed to act as a governing force (e.g., collecting taxes, arresting people, patrolling with visible weapons). Such activities undermine the legitimacy and credibility of the local government the United States is supporting and present the adversary force as an alternative. U.S. forces may not always be able to prevent adversary forces from moving covertly, but they can readily detect and stop overt operations (roadblocks, weapons displays, vehicle convoys).

Not confined to peace operations only, detecting and stopping combatant movement may be necessary in noncombatant evacuation operations, hostage rescues, and other special operations. A traditional approach to this problem would put friendly ground forces throughout the city, manning observation posts and roadblocks, patrolling, and generally making it difficult for adversary forces to move without detection.

Both ground sensor networks and airborne platforms have the potential to reduce the manpower demands and risks to friendly forces associated with these operations. Although friendly ground forces are likely to still be required, they could be more effective if cued to problem areas by unmanned sensors. In some cases, such as a NEO or hostage rescue, airborne platforms (e.g., an AC-130) might operate independently to detect and interdict hostile forces. In Bosnia, JSTARS aircraft accomplished the surveillance task in rural areas by ensuring that the Bosnian, Serb, and Croatian forces lived up to the terms of the Dayton Agreement and kept their armored and other combat vehicles in holding areas. In some urban situations, JSTARS may have a role to play also; however, in urban core areas, building shadows and the commingling of adversary vehicles with civilian traffic (e.g., as Somali "technicals"[2] were) will make it difficult to monitor vehicle movement, and JSTARS is unlikely to be able to distinguish between them. In our judgment, a distributed sensor system

---

[2]*Technicals* were pickup trucks and jeeps with machine guns mounted on the back.

will be necessary to detect combatant movement under these conditions.

A distributed sensor system could combine imaging and non-imaging sensors on both air and ground platforms. The imaging sensors would look for military vehicles, "technicals," and other, more-overt displays of weapons. Non-imaging sensors would look for hidden weapons or explosives or use pattern analysis to detect anomalous vehicle or personnel movements. Initially, these sensors could supplement ground-force monitoring in the more dangerous or problematic parts of the city.

If unmanned ground sensors prove to be effective and some of the more exotic technologies become cheap and reliable, sensors could be placed throughout much of the city. These sensors would alert a weapon controller, who would then direct airborne sensors or ground patrols to take a closer look. When adversary ground forces could be clearly identified (e.g., massing at a militia leader's compound as they did in Mogadishu), airborne strike platforms could be used to destroy them; or ground forces could be alerted to investigate further.

In this more-ambitious vision, the role of ground forces shifts to that of quick-reaction force for situations where a person on the ground is needed because of problems with target identification or because combatants are too intermingled with noncombatants for attack with standoff weapons. This concept might allow for a reduction in the number of required ground forces; if a small number were deployed, they would need some means to move quickly to the site of the problem. One advantage that ground forces have over most airborne platforms is their ability to detect and fire on fleeting targets. The longer the time of flight of the weapon, the harder it is to engage targets moving quickly in and out of cover.

Figure 5.1 illustrates the basic concept. On the left, a UAV with an EO/IR camera detects an insurgent roadblock and relays this information to the controller. The controller identifies the personnel as hostile and directs an AC-130 to take them under fire. On the right side of the figure, a "ground sensor" (located on the top of a building)

RAND *MR1187-5.1*

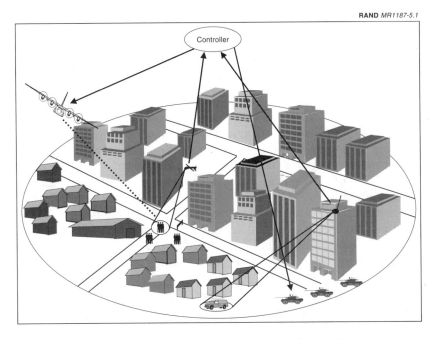

**Figure 5.1—Sensors Detect and Stop Combatant Force
Movement or Operations**

equipped with radio-frequency (RF) resonance-detection equipment
detects what appears to be a load of assault rifles in a parked van. In
this case, friendly ground forces (in the armored vehicles on the
right) are sent to investigate.

PROVIDE RAPID, HIGH-RESOLUTION IMAGERY
FOR TARGET ID

As the USAF discovered in Kosovo, from medium altitudes it is often
impossible to distinguish between various types of vehicles or between adversary and friendly dismounted personnel. To aid in situations in which ground observers are not available to assist in target
ID, it would be enormously useful if combat aircraft (whether F-16s
or AC-130s) had access to offboard sensors flying at lower altitudes. If
Predator or other UAVs are on-station, they may be available for this
role. However, the limited number of such platforms and concern

about their survivability at lower altitudes may mean that combat aircraft cannot always count on UAV assistance.

In addition to these surveillance platforms, we propose the development of a set of air-launched disposable sensors for target ID in urgent situations. It would be possible for the parent platform, either a fighter, AC-130, or larger UAV, to drop a miniature glider with a small, low-light-level TV camera, GPS receiver, auto-pilot, and data link. The mini-glider could be folded into a container until deployed. However, a system like this would take many minutes to glide down from 15,000 or 20,000 ft to low altitudes—less than 500 ft—where its camera could provide the high-resolution images necessary for target identification. During that time, many things could happen. For example, the target could move away or, in an urban environment, move into a building. Therefore, it is important that any disposable identification sensor arrive at low altitude near the target very quickly.

One possibility would be for an F-16 or other fighter aircraft to carry a mini-UAV in a canister on one of the wing weapon stations. When the F-16 detected a situation requiring target ID and some endurance was desired, the UAV canister would be released and fall quickly to lower altitudes. A drogue chute would slow the canister and, at a few thousand feet altitude, a folding-wing UAV would be deployed. That UAV would fly at low altitudes, providing imagery back to the friendly aircraft for up to a few hours to allow confident target identification, attack, and battlefield damage assessment (BDA) or, as in the Kosovo example, to avoid accidental attacks on civilians. This kind of endurance over the target would often exceed that of the fighter deploying the UAV and would allow a UAV deployed by one fighter to provide imagery to its replacement on-station.

Another possibility is to build a container for the sensor package. The container would be shaped like a lifting body, with small control surfaces.[3] Once dropped, the container would fall rapidly to low alti-

---

[3]RAND colleague Randall Steeb points out that this is similar in concept to the video-imaging round that the U.S. Army experimented with a few years ago. One version used the spiraling of the shell to produce a scanned image; another version used a parachute-dropped, GPS-equipped camera.

tude, but would generate enough lift to allow it to maneuver to co-ordinates anywhere within a few miles of the launch aircraft. However, a shape that is well adapted to rapidly maneuvering to a given set of GPS coordinates at low altitude is poorly suited for taking even a brief close-in look at whatever may be of interest there. It would not be capable of maintaining level flight or circling a target at close range without rapidly depleting its airspeed, which would re-sult in an almost equally rapid aerodynamic stall and crash.

Therefore, we envision the lifting body deploying a small, control-lable parafoil as it passes about 1,000 ft AGL. By the time it reaches 500 ft AGL, the parafoil should be fully unfurled, slowing the con-tainer. At about this altitude, it would begin a slow spiral around the GPS coordinates of interest. The diameter of the circle might be se-lectable before launch to allow the operator to tailor the resulting view, somewhat, to the suspected target and environmental condi-tions. Something as simple as "Near," "Medium," and "Far" circle di-ameters (determined through empirical tests during sensor devel-opment) might be sufficient.

The container could be suspended so that the camera looks to the inside of the descending turn. In its simplest form, a system like this would just continue to fly descending circles about the target coordi-nates, sending back images to its parent craft all the while, until it hit the ground. More-sophisticated versions might allow operators to steer the parafoil to change the camera's azimuth view and include some means of controlling camera elevation view. These more complicated schemes would probably require a human to control the expendable camera once the parafoil deployed. This might make them more appropriate for multi-place (i.e., two or more crew-members) aircraft or large UAVs, such as F-15Es, AC-130s, or Predator, so that crewmembers or operators can dedicate their full attention to operating the expendable camera for a short time. However, the simpler autonomous version would still be a valuable target-identification tool for single-seat fighter or attack aircraft operating at medium altitude.

In Figure 5.2, we illustrate how this GPS-guided parafoil might be used to accomplish a key operational task. In this concept, we envi-sion an AC-130 or other combat aircraft patrolling an urban envi-ronment. The AC-130 detects a suspicious gathering of vehicles and

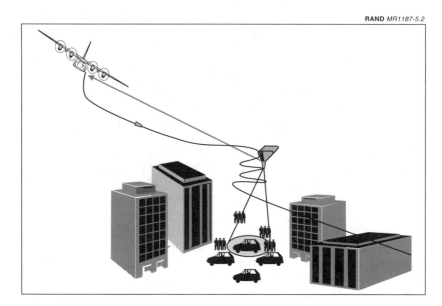

RAND *MR1187-5.2*

Figure 5.2—GPS-Guided Video Camera Is Used for Target ID

personnel but cannot positively identify them as hostile. Strict ROE prohibit engaging them without positive ID, and no friendly ground forces are available to investigate. Instead, the AC-130 crew deploys the lifting-body-shaped camera that flies to GPS coordinates programmed by the crew.

High-resolution low-light TV or IR images are relayed back to the AC-130. The parafoil can spiral all the way down onto hostile forces or can carry the camera clear of them. Either way, very high-resolution images should be obtainable as the parafoil passes below 500 ft, perhaps passing tens of feet over the target. The AC-130 by this point has made the determination that the people are in fact hostile forces being resupplied and attacks them with 40mm or 105mm rounds.

Such a capability would be enormously valuable in a host of situations in both rural and urban terrain. As advanced MANPADS proliferate around the world, it is likely that manned platforms will increasingly avoid loitering at lower and medium altitudes near threat forces. Air-deployable offboard sensors would give combat aircraft all the surveillance advantages of loitering with none of the attendant risks and many of the capabilities of more-expensive surveillance platforms in a package they control.

DETECT AND NEUTRALIZE ADVERSARY AMBUSH POSITIONS

The ambush is a common technique that adversary forces are likely to use against friendly forces. Although their size and the type of weapon used can vary, classic infantry ambushes involve a small unit (squad or platoon) surreptitiously taking up hidden positions (often under cover of darkness) and waiting for an adversary patrol, convoy, civilian vehicles, or other targets to enter the kill zone. The ambushers use assault rifles, machine guns, antitank weapons, grenades, and mines to create a horrific volume of fire, often at very close range, against the victims. Careful siting of weapons is used to put fire down and across the road or trail so that there is no cover for the ambushed force. The suddenness and volume of fire can wipe out a patrol in seconds. The ambushers may take weapons and intelligence materials off the bodies or simply cease fire and leave.

If the ambush is badly sited or executed, adversary forces may be able to return fire and maneuver against the ambushers, particularly in ambushes against forces that are too large to be contained entirely within the kill zone. In those cases, the forces not in the kill zone are likely to maneuver around the flanks of the ambushers to bring fire on their positions and to cut off their escape. Also, nearby forces may come to the aid of those caught in the ambush. To avoid these dangers, ambushers do not linger. A well-executed ambush may be over in under a minute and rarely extends to more than a few minutes.

Weak forces may use ambushes to maintain the initiative, control territory, demoralize adversary forces, or gather intelligence (from maps, radio codes, and other operational information taken from the bodies of ambush victims). In an urban setting, adversaries may use ambushes to inflict heavy casualties and, thereby, so intimidate friendly forces that they are afraid to patrol particular sectors of the city. These areas could then come under the *de facto* control of, for example, insurgent forces or criminal elements. Alternatively, adversaries may seek to inflict casualties, on a peacekeeping force for example, as a way of undermining international support for the intervention. Since patrolling is essential in many operations, U.S. forces need to find ways to better protect small units from ambushes.

Ideally, we would like to be able to detect and neutralize ambushes from standoff. As the above discussion illustrated, there are four components to an ambush: movement to the site, hiding at the site, the ambush itself, and escape. In urban operations, the ambushers may move to the site overtly; however, in the kind of operations we envision predominating, they probably will move in a more covert way. Thus, visual observation will probably not allow them to be singled out from the background civilian traffic. One exception would be if they attempt to move covertly at night but are detected by airborne or ground-based low-light TV or IR sensors. Even if their weapons were hidden, their movement could stand out if the streets were deserted and they were moving in rushes to minimize visibility from street observers. Also, environment-shaping measures such as curfews can help by making *any* activity at certain times or places suspicious. Alternatively, unmanned ground sensors might detect the ambusher's body heat, weapons, or explosives if the ambushers passed by them. This would require either a large network of sensors or careful siting to maximize the probability of the ambushers passing by the sensors.[4]

A more difficult problem is detecting the ambushers once they are established in their attack positions. Airborne or ground-sensor platforms may be able to detect some ambush positions on streets, balconies, or rooftops, perhaps even inside of structures. Once detected, they could be attacked using traditional air or ground weapons or some of the more-specialized munitions discussed later in this chapter. Generally, detection of ambush positions will require fairly high-resolution IR or visual sensors, although other sensor types, such as radars that can detect weapons, might play a role. It is unlikely that airborne platforms *randomly* observing the urban land-

---

[4]In many situations, particularly those relating to defense of key installations, it will be possible to site ground sensors at choke points through which adversary forces would have to pass. For example, USAF security police described to one of our researchers a few years ago how security forces were able to use ground sensors to count adversary personnel. During air base defense exercises in Korea, security forces were able to put ground sensors on likely aggressor avenues of approach. As the aggressors passed by in file, USAF security forces at the remote monitoring site were able to count them and direct a quick-reaction force to the scene.

scape will detect ambushes, because either their sensor resolution will be too low (in the case of higher-flying aircraft with high coverage rates) or the coverage rate will be too slow (in the case of lower-flying aircraft taking close looks at individual streets and structures).

For this reason, ambush detection will need to be focused on specific streets or buildings for specific periods of time. For example, each foot or vehicle patrol could have an unmanned sensor platform (ground or air) precede it in search of suspicious activity, which would focus the sensor on the area of primary concern to the patrol and at the critical time. Such a platform could be controlled either by the patrol leader or at a rear command post. To the extent that more-exotic sensors such as chemical sniffers (to detect ammunition and explosives), weapon-detecting radars, etc. become feasible, this platform might be able to detect the actual ambush positions. The positions might also be detectable by IR, visual, or multispectral sensors.

If the ambushers are hiding inside of buildings (probably just below or beside windows), they will be very difficult to detect, requiring more-intrusive approaches. These approaches might include rifle-launched sensors that would fly through a window and attach to a wall, micro-UAVs doing essentially the same thing, or perhaps sensors that could be rolled or thrown into a room.

Sometimes, adversary forces will be able to set up an urban ambush without local civilians detecting it; other times, locals will be aware of the ambush because it is being set up in the building they live or work in. To the extent that local civilians are at odds with the adversary force, they may warn authorities of the ambush. Even when the locals are supporters, perhaps even involved in helping set up the ambush, they may take steps to protect themselves or family members, or otherwise engage in nonroutine behavior that can be observed.

An experienced operator (either a member of the patrol or rear-area sensor operator) may detect changes in the social and physical landscape that suggest there is something wrong, although he cannot see the actual ambushers. For example, an infantry patrol leader might notice changes in people's behavior, number or location of vehicles, whether shop windows or doors are open or closed, presence or absence of children, dogs barking, unusual silence, or other signs that would indicate danger. The more experience the patrol has with a particular patrol route, the more likely they will be to detect such changes. Many combat infantrymen and policemen have reported detecting such anomalies and being saved from imminent attack, sometimes without being conscious of what exactly is wrong with the picture. Unfortunately, these signs are not always clear and the patrol may not detect them until it is too late.

Rather, what we need is a way to observe these patterns safely from standoff and in a more systematic and reliable way. The UAVs discussed above could be used to extend the vision of the patrol or other observer. Using experience or a database of photos taken at similar times and days of the week, the patrol or observer could compare the current picture with the baseline. There is some danger that the appearance of an unmanned platform would be recognized by adversaries as a precursor to the arrival of the patrol, but this seems a small price to pay for the increased security. Also, there are a variety of techniques that could be used to make the UAV more covert.

An alternative method (illustrated in Figure 5.3) that has utility beyond this particular task would be to use a network of ground sensors to continuously observe locations of interest, which might include major roads, marketplaces, town squares, trouble spots, and likely ambush locations on patrol routes. The network could use IR or visual sensors, but unless some technique were developed to automatically analyze the images, this approach would be feasible only for small networks. If, however, seismic, acoustic, sniffing, magnetic, and other sensors could be developed, it might be possible to rely more heavily on computers to do pattern analysis. For example, automated analysis of vehicle- or foot-traffic patterns might be possible. These or other sensors might be used to alert sensor operators or intelligence analysts to anomalies. Operators and analysts could then use imagery to delve deeper into the mystery, comparing stored images from similar days of the week and times.

From a force-protection perspective, it would be best to have both the wide-, or at least wider, area surveillance associated with the ground sensors and a small UAV (perhaps even multiple UAVs) providing imagery and other data directly to the patrol.

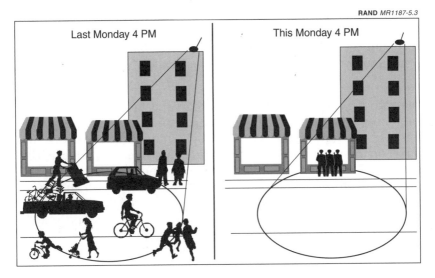

Figure 5.3—Pattern Analysis Detects Anomalies, Warns of Ambushes

DETECT AND NEUTRALIZE SNIPERS

Snipers are a perennial problem in urban military operations. Threatening civilians and military personnel alike, snipers can shut down civil activities, hurt morale, cause politically significant casualties, perhaps even stop a NEO from an urban embassy. Snipers armed with the increasingly popular .50-caliber rifle (using incendiary or armor-piercing bullets) can threaten lightly armored vehicles and hovering or landing helicopters.

For our purposes, a *sniper* is a single combatant (sometimes teamed with an observer) who selectively engages targets from a location that offers superior fields of view. He may be a specially trained sniper, a regular infantryman, or a civilian irregular.

An untrained sniper may fire from a balcony, rooftop, or window with a standard assault rifle such as an AK-47, using either the weapon's iron sight or a scope. Most sniper casualties in recent civil conflicts (e.g., in Bosnia and Somalia) were probably the work of untrained snipers, either irregulars or soldiers with only basic marksmanship training. U.S. forces encountering such snipers during operations in Somalia found that countersniper teams flying in helicopters or on the ground were able to routinely detect and dispatch these threats because the snipers failed to use cover or tended to fire too many times from one location.[5]

In contrast, professional snipers rarely fire from open positions.[6] Instead, they will covertly enter a hide and take substantial time preparing their firing site. In an urban setting, a well-trained sniper will use windows, small holes through shingles or walls, or other small openings for observation and fire. They often will build a firing platform in the back of a room hidden by netting.[7] Although this cover limits their field of view, it also masks muzzle flash and makes them extremely difficult to detect. Professional snipers typically use special bolt action or semi-automatic rifles,[8] scopes and match-grade ammunition, spotting scopes, and camouflage (e.g., "Ghillie suits"), and they are trained in ballistic calculations and other tradecraft.

The U.S. practice, for both police and military snipers, is for snipers to operate in pairs, rotating between shooter and observer duties. The observer uses a spotting scope to help adjust fire and provides security. He typically is armed with a semi-automatic weapon. Adversary snipers may operate as individuals or in pairs. Finally, the professional sniper will take only a few shots from his position before

---

[5]Tony Capaccio, "U.S. Snipers Enforced Peace Through Gun Barrels," *Defense Week*, January 31, 1994, p. 1.

[6]What constitutes an *open position* will vary with the situation. For example, a police sniper on a rooftop, while visible to airborne platforms, typically is only concerned with remaining hidden from his criminal target in a building or at street level. In contrast, in a combat situation, a sniper going against U.S. forces would have to worry about detection from airborne platforms. Even a professional sniper might fire from the open under these conditions if the mission required it. The point here is that U.S. forces should not count on snipers routinely being detectable through simple visual observation of open spaces.

[7]See U.S. Department of the Army, *An Infantryman's Guide to Combat in Built-Up Areas*, Washington, D.C.: FM 90-10-1, October 1995, especially pp. 5-23–5-34 and Appendix E for details of firing platforms and positions for infantry in structures. Appendix J covers countersniper techniques.

[8]Sniper rifles include the Russian SVD and Mosin-Nagant, the U.S. Remington 700 and McMillan .50-caliber, and the British Parker-Hale M85. However, trained snipers have been quite effective with less-sophisticated weapons. For example, Irish Republican Army (IRA) snipers have been very successful over the years using AR-15s with basic scopes.

leaving the area or moving to an alternate firing position.[9] For this reason, a sniper should be considered a time-critical target.

There are three situations in which U.S. forces are likely to need to counter snipers:

- The first is in the defense of U.S. or allied facilities or ground-force positions. U.S. embassies, air bases, airports, ports, barracks, allied governmental buildings, television, or other public facilities might come under attack from snipers. At best, such attacks are a nuisance; at worst, they have the potential to cause serious loss of life or damage to high-value assets such as parked aircraft.

- A second situation is when friendly patrols come under sniper fire. Routine sniper attacks on patrols could seriously disrupt their ability to interact with the local populace, observe activities, and collect intelligence. Force-protection concerns could limit patrol frequency, locations, and duration; undermine morale; and cause the patrols to be so defensive that they were ineffective in their primary mission.

- Third, civilian populations can be harassed and intimidated through sniper attacks on foot traffic, marketplaces, parks, and other places where civilians congregate. For example, during the Bosnian civil war, Serbs in the suburban hills surrounding Sarajevo routinely fired into the center of Sarajevo, particularly down "Sniper's Alley," a road near the Holiday Inn.

U.S. forces have traditionally used their own snipers to stalk and kill adversary snipers.[10] Although manpower-intensive and time-consuming, this is an effective way to counter professional snipers. U.S. countersniper teams are even more effective against untrained snipers but can quickly be overwhelmed by sheer numbers. These teams are simply too few to effectively counter irregular and other

---

[9]For a thorough treatment of sniper equipment, techniques, and tactics, see John L. Plaster, *The Ultimate Sniper: An Advanced Training Manual for Military and Police Snipers,* Boulder, Colo.: Paladin Press, 1993.

[10]See Plaster, 1993, pp. 365–394, for a discussion of countersniper tactics.

forces deploying many infantrymen or even untrained militias as snipers throughout a city.

To supplement infantry countersniper teams and allow them to focus on priority missions, DoD has been exploring other countersniper concepts. Acoustic, radar, passive IR, and scanning lasers have all been tested for their applicability against snipers. The Stingray scanning laser system on the Bradley Fighting Vehicle, for example, can be used to detect sniper optics (telescopes or night-vision devices) and alert the gunner; in automatic mode, it can engage and neutralize optics.[11]

We propose a two-track approach for expanding countersniper operations in the urban environment, as illustrated in Figure 5.4.

For fixed facilities and known problem areas, unmanned ground sensors would be deployed by ground forces or air.[12] Scanning lasers, acoustic arrays, and passive IR systems all hold promise for this mission. Since scanning lasers can potentially detect the sniper before he has fired, they should be used wherever possible; the other approaches, particularly passive IR, should be used in combination with scanning lasers to increase the probability of detecting and destroying sniper threats after they have fired.

In our concept, we use an unmanned sensor equipped with a scanning laser, laser designator,[13] and EO/IR camera. When the scanning laser detects optics, the camera would automatically slew to that location and a controller would be alerted and automatically provided with three-dimensional (3-D) coordinates for the

---

[11]U.S. Department of the Army, FM 90-10-1, 1995, p. J-8.

[12]An important shortcoming is the current inability to implant urban ground sensors from the air. Past aerial-delivery sensors were either high-speed spikes that implanted themselves in the ground (primarily seismic sensors) or acoustic sensors dropped by parachute. These approaches are viable for operations in undeveloped areas but have limited utility in urban settings. Rather, what is needed are small, difficult-to-detect sensors that can be covertly implanted on building tops or sides. One concept that we recommend exploring would use small VTOL UAVs to implant such sensors.

[13]A *laser designator* is a device that illuminates a target with laser energy so that a weapon equipped with a laser receiver can guide in on the beam.

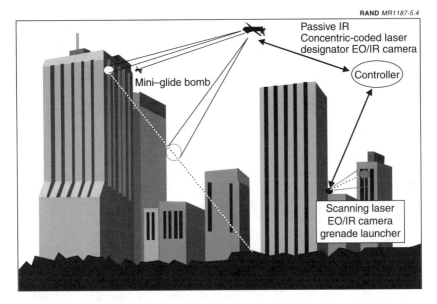

Figure 5.4—Armed UAV and Unmanned Ground Sensors Counter Snipers

location.[14] Automated decision support software would compare these coordinates to a 3-D database to determine whether the location was known to house friendlies or noncombatants. Using this information, what he knows about friendly operations, and what he sees remotely through the camera, the controller would then make a determination on the next step. If it appears to be a legitimate target, the controller could send ground forces to the location or preemptively use nonlethal or lethal weapons against the site. We envision equipping at least some of the sensors with a grenade launcher. The launcher would fire laser-guided grenades with sufficient accuracy to go through an average-sized window and have sufficient lethality to take out a sniper without harming noncombatants in adjacent rooms or on adjacent floors.

---

[14]All of our concepts assume that the USAF will acquire the ability to do near-real-time three-dimensional imaging of urban environments and that this imaging will be used to produce a three-dimensional coordinate system for navigation, battle management, and weapon-system guidance.

To deter random sniper attacks on civilians or attacks on friendly patrols, we propose supplementing the fixed sensors with a passive IR system like Lawrence Livermore's Lifeguard system, on a UAV. The UAV would also be equipped with an EO/IR camera, concentric-coded laser[15, 16] designator, and mini–glide bombs with small warheads. When the passive IR sensor detects a hot bullet against the cooler background of the air, it uses a ballistic model to backtrack to the firing location. Separate calculations are done for each bullet fired, enabling the sniper's location to be determined with sufficient accuracy for counterfire. When used in the fixed ground mode in the line of fire, the Lifeguard system slews a camera to the sniper location, allowing a friendly sniper or other weapon operator to engage.

For an airborne platform it is possible that the sensor would have line of sight to the bullet in flight, but not to the firing location, which might be blocked by another building. Thus, slewing a camera to the sniper will not always be possible. Rather, the UAV fire-control computer would need access to the 3-D city database so that the ballistic track could be compared and the likely firing location determined.[17] If it were not already in a position with line of sight to the sniper's location, the UAV would maneuver so that it was. The EO/IR camera

---

[15]A circular laser puts energy around the window but not on it. The weapons would be programmed to fly through the middle of the laser circle. Concentric-coded lasers would put several laser rings around the target, using different frequencies of lasers to convey information about target location to the weapon.

[16]As noted in Chapter Four, one problem with laser designation in urban environments is that the laser energy may reflect off window glass or, where windows are missing, go into a structure but not reflect enough energy out to guide a weapon. One possible solution would be to put laser energy on less-reflective surfaces around the window, perhaps in concentric circles. The simplest near-term concentric coding might involve a purely spatial code that could be traced using technology similar to that employed in laser light shows. For example, the pattern closest to the target might be two nearly concentric circles with slightly offset centers, so that the distance between them appears to be larger at the top of the circles than at the bottom. As the seeker scans across the pattern, the amount of separation would indicate whether the seeker is approaching high or low but, also, by virtue of the number of circles, would indicate the distance to the target. In the future, a more advanced laser could be used to provide similar data by varying the laser frequency or by modulating the laser signal.

[17]A simple algorithm could rule out interior spaces and windows on the far sides of buildings, identifying the most likely firing position along the ballistic track.

would then be directed at the sniper location so that the controller could put eyes on target before releasing a weapon.

At this point, a mini–glide bomb[18] would be released, flying a course over and between buildings using enhanced GPS signals and the 3-D database to navigate to a position in front of the target building (see Figures 5.5 and 5.6). A high-flying UAV would act as a GPS pseudolite,[19] increasing the accuracy of the signal and rebroadcasting it at a frequency that can be received in the urban canyons.[20] The concentric-coded laser designator on the mother UAV would illuminate the

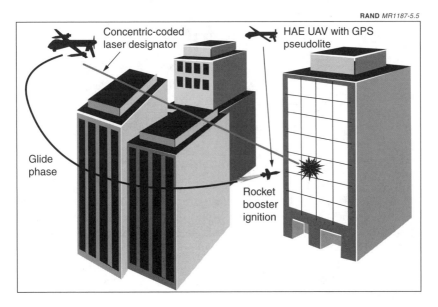

RAND MR1187-5.5

Figure 5.5—Mini–Glide Bomb Flyout

[18]Other possibilities would be to use ground-based systems such as optically guided Enhanced Fiber-Optic Guided Missile (EFOG-M) or perhaps maneuvering mortar rounds.

[19]Pseudolites are ground-based or airborne transmitters that supplement or replace GPS for navigational purposes. See the discussion in Chapter Six.

[20]See Chapter Six for a fuller discussion of GPS use in urban settings.

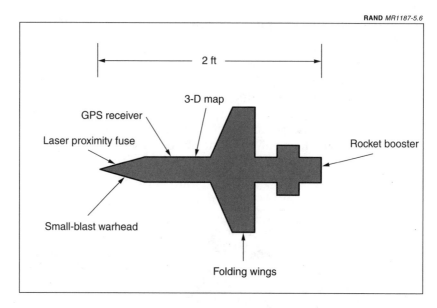

RAND *MR1187-5.6*

Figure 5.6—Schematic of Mini–Glide Bomb

target, providing terminal guidance for the glide bomb. The glide bomb would maneuver and ignite a small rocket motor to give it sufficient energy to penetrate a window or other minimal obstruction if necessary. A laser proximity fuse would then detonate the small warhead once it had entered the sniper-occupied room.[21] By using a small, light platform, flying a level course, and bursting a small warhead inside of the room, this concept should allow for effective counterfire without putting noncombatants at risk in adjacent rooms or on adjacent floors.

MONITOR HIGH-PRIORITY TARGETS

In addition to more-general surveillance requirements, urban operations—particularly counterterrorist, counterdrug or WMD-related—may require continuous monitoring of a building or fairly

---

[21]We envision not only a lethal fragmenting warhead of roughly grenade size, but also incapacitating gas, a stun grenade, or other nonlethal weapons as other possibilities.

small area. There may be a need to observe or listen to activities in a particular room; to monitor personnel, equipment, or vehicles entering or leaving a building; or to otherwise observe activities at a town square, park, or other fixed site.

Most of the time, AC-130s, Predators, and other existing platforms using EO, IR, or radar sensors have sufficient resolution to accomplish these missions covertly[22] from medium to high altitudes. If, however, the mission requires identifying a particular person, small piece of equipment, or small package entering or leaving a building, imagery equivalent to that provided by a police stakeout squad in a vehicle or nearby building would be necessary. Larger platforms operating at standoff distances do not have sufficient resolution to accomplish these extremely demanding tasks. In the following discussion, we explore the possibility that low-flying UAVs or unattended ground sensors could achieve this very high level of resolution.

Mini- and micro-UAVs (with wingspans from 8 ft or so down to bird size) have much utility in urban settings. They can fly down into urban canyons, thereby gaining excellent viewing angles through windows and of streets, alleyways, and other narrow passageways. However, it does not appear that they can conduct enduring covert surveillance with EO or IR sensors. The problem is the mismatch between what the UAV needs to do to monitor the site and what a human observer at the site needs to do to detect the UAV. The UAV sensor would need resolution on the order of inches to identify specific individuals or very small packages. To get this resolution requires that the UAV either carry a camera with a long-focal-length lens or get very close to the target.

In exploring various combinations of UAV size, associated payload, and sensor range, we could not find one that would allow the UAV to get sufficiently close to identify a human and still remain undetectable to the adversary observer. For example, a typical slow-flying UAV with a wingspan of 8 ft can carry roughly a 5-lb payload. A standard 5-lb optical-sensor package has sufficient resolution to identify

---

[22]That is, they would be difficult or impossible to detect with the naked eye. If the adversary had radar coverage or advanced IR systems, these aircraft would be detectable.

a specific person (e.g., Osama bin Laden as opposed to Saddam Hussein) in daylight at a distance of about 1,400 ft.[23]

At this distance, not only could the UAV be spotted but it could easily be shot down. The adversary human observer needs only to detect the UAV and identify it as an aircraft, and the UAV is moving, which makes detection much easier. Resolution of 1 ft will probably be adequate to determine that the UAV is not a bird. To make matters worse, the UAV can be detected acoustically, and it is difficult to make them very quiet. Whether flying a racetrack offset pattern or an orbit around the surveillance target, the regularity of the movement would make the UAV stand out as a man-made object after only an orbit or two.

Alternatively, we could use micro-UAVs, insectoids, or a collection of ground sensors on nearby buildings to avoid detection. Micro-UAVs or insectoids (either flying, hopping, or crawling) would use their small size to get extremely close to or inside a target building. Once on or in the building, they might attach themselves to a wall and observe. Such sensors and platforms are being explored at Los Alamos and other laboratories, but a host of aerodynamic, power-supply, navigation, and communication challenges need to be overcome before they have much operational utility. These systems offer promise for some high-priority, specialized surveillance missions; however we view them as being unlikely to be practical for routine surveillance missions in the near term.

The only enduring, high-resolution covert sensor that is practical in the near term is either an unmanned ground sensor or a ground observation team. Even with these alternatives, there is some chance of discovery upon insertion or at some later point. Inserting ground sensors covertly is tricky and requires either ground personnel (perhaps disguised as maintenance workers) or a precise and quiet airborne mode. Also in cases where enduring surveillance is required, ground sensors may fail and have to be replaced and ground observers must be rotated or resupplied.

---

[23]See pages 172–173 for more on these performance trade-offs.

In our concept, we envision deploying miniaturized ground sensors by VTOL UAV.[24] These would be deployed at night by a small, quiet VTOL UAV. To avoid detection, the UAV would need to fly a profile that maximized masking by buildings or rooftop structures. For example, most multistory urban buildings do not have completely flat roofs. Typically, the roof space also contains a 1-story structure housing elevator, heating, or other machinery. A VTOL UAV could land on the far side of this structure to escape visual detection from the target while implanting the sensor, but doing so would put the sensor in an undesirable location. For this reason, the sensor would need some limited mobility so that it could crawl to the correct location on the roof. Once it was in position, a small telescoping arm would raise the optics or other sensor above the roof lip for viewing, minimizing its signature from the target building (see Figure 5.7). To minimize accidental discovery by people who might have occasion

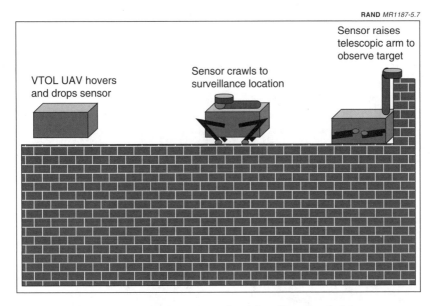

RAND *MR1187-5.7*

Sensor raises telescopic arm to observe target

Sensor crawls to surveillance location

VTOL UAV hovers and drops sensor

Figure 5.7—Covert Placement of Rooftop Unattended Sensor

---

[24]These would range in size from a shoebox down to a large coin.

to visit the rooftop, the sensor housing would need to be designed to blend in with the surroundings, masquerading as a piece of electrical equipment or other cultural artifact. Alternatively, sensors might be designed for emplacement on walls or other hard-to-reach places. In these cases, they also would need to be designed to blend in with the surroundings.

As Figure 5.8 illustrates, sensors on multiple buildings would provide continuous surveillance of all building entrances, selected interior spaces with windows, rooftops, and balconies as appropriate. In addition to surveilling this fixed site, it might also be necessary to follow a person or vehicle after he or it had left the site. We envision a VTOL UAV for this task. It would land on an isolated location on a nearby rooftop (perhaps on top of the machinery structure), where it would stand by. If a target person or vehicle left the structure, this UAV would take off and follow it. For this concept to work, the UAV would have to be partially powered up and able to achieve flight within a short period of time, most likely under a minute. To give the UAV a

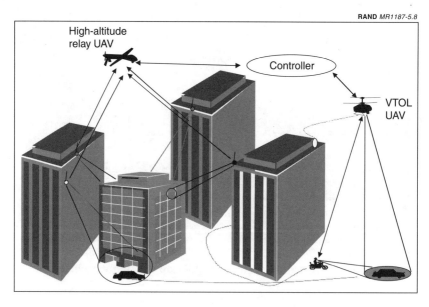

Figure 5.8—Unattended Ground Sensors, SOF Team, and UAVs
Monitor High-Priority Target (Covert)

bit more time to get airborne, it might be necessary to implant additional ground sensors along avenues of approach to the building. Another possibility would use a medium-altitude UAV to provide interim coverage until the VTOL UAV was able to take over, although lines of sight to the target would be sporadic in urban-core areas, particularly if the target made many turns.

The most robust concept would use some combination of ground personnel (on foot and in vehicles), unmanned ground sensors, and airborne platforms. USAF and joint exercises (as well as many years of law-enforcement experience) using airborne platforms and ground personnel have shown this to be a highly effective way to covertly follow vehicles.

RESUPPLY ISOLATED FRIENDLY GROUND FORCES

During urban operations, friendly ground forces may become isolated and need resupply from the air. We specify urban operations, because isolation is much more likely to happen in urban settings because of the difficulty of preventing adversary infiltration of friendly lines. Also, friendly patrols or special operations forces often will be operating in contested or adversary-controlled terrain. In many cases, adversary roadblocks, downed bridges, or rubble in roads will mean that such forces cannot be resupplied via ground routes.

This is exactly what happened to Task Force Ranger in Mogadishu when the operation went awry. Unable to withdraw or be reinforced by ground, the task force—in desperate need of ammunition, intravenous fluid bags (IVs), water, and other supplies—found shelter in a few small buildings. One helicopter did manage to hover over one friendly group and drop some supplies, but it was so badly shot up in the process that it barely made it back to the airport. No other attempts were made.[25]

In such situations, adversary fire will prevent helicopter resupply, and traditional fixed-wing airdrop from safer altitudes lacks the precision needed to put the supplies in the right hands.

---

[25]See Bowden, 1999, pp. 230–231.

In the concept illustrated in Figure 5.9, we propose developing GPS-guided resupply canisters. The isolated unit would transmit its supply request and GPS coordinates to a control center, which would dispatch an aircraft—fighter, transport, or rotary wing. A canister could be prepackaged with basic supplies or tailored to support specific mission needs. It could be released from a variety of altitudes and standoff ranges, depending on the local situation. The canister would be aerodynamically shaped, have control surfaces similar to a GPS- or laser-guided bomb, and would be programmed to fly to the GPS coordinates of the isolated unit. At a relatively low altitude (to prevent drift), a drogue chute would be deployed to slow the canister. Shortly before impact, airbags also would deploy to cushion the impact.

Basic engineering and field tests will have to be done to determine the feasibility of this concept. However, the combination of shock-protected compartments, air bags, and a drogue chute should allow precision resupply in urban settings.

RAND *MR1187-5.9*

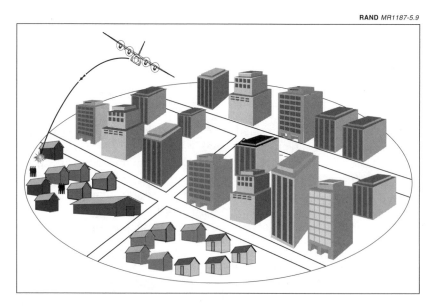

Figure 5.9—GPS-Guided Canister Resupplies Friendly Forces

PROVIDE CLOSE SUPPORT FOR GROUND FORCES

As noted above, it is easy for small ground forces to become isolated in urban settings. Urban structures limit both visibility and fields of fire, horizontally *and* vertically, making it difficult for ground forces to provide mutually supporting fire. Traditional fire support from artillery is often limited in urban areas because of its low-angle trajectory. Mortars, which fire at much steeper angles, are better able to get over buildings. However, both mortars and artillery are insufficiently accurate to use in situations where collateral damage must be minimized. Air forces can provide immediate and accurate fire support to friendly ground forces engaged in close combat.

Figure 5.10 illustrates one concept for providing fire support to a small ground element. In the illustration, a friendly patrol at ground level has become pinned down by adversary forces firing from fourth-story windows across the street. The friendly force uses a laser rangefinder/GPS receiver like the Viper system to determine the GPS coordinates of the adversary force. They also use a circular or

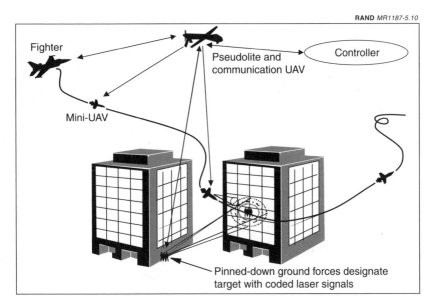

Figure 5.10—Attack Mini-UAV Provides Close Support

concentric-coded laser designator to illuminate the adversary force. The GPS coordinates of the adversary force are relayed, along with their fire-support request, to a friendly command facility. The controller enters the GPS coordinates into a 3-D map/urban database and receives basic information about the building. From this, the controller learns that the adversary force is firing from an apartment building occupied by noncombatants. Under these conditions, strict ROE must be observed to minimize civilian casualties. A controller directs a friendly aircraft equipped with armed mini-UAVs to provide fire support.

The aircraft—in this case, a fighter—releases the UAV, which uses GPS signals to fly toward the adversary's position. At this point, its onboard guidance system determines the best approach route, using an onboard 3-D map to negotiate the city streets, and the UAV detects the laser reflection off the adversary's position and flies a path directly in front of it, firing multiple grenade-sized explosives or perhaps a nonlethal incapacitating agent through the windows. The UAV would have a multiple-shot capacity and could return to fire again if necessary. Although such a limited-effects weapon might not disable or kill all the adversary combatants, it would probably produce sufficient shock to allow the friendly forces to escape.

Ideally, such weapons would carry variable-effect munitions, which allow the amount of explosive power to be adjusted for each mission. Again, the technical details of such a weapon have yet to be worked out, but the concept is fundamentally practicable.

To summarize, the key characteristics of this weapon are its small size and slow speed. Both characteristics enable it to maneuver in the urban canyons and to either fly by a window, firing a weapon as it passes, or to turn and fly through the window and detonate inside. The small size and weight and the slow speed of this weapon would minimize penetration and collateral damage in civilian structures.

THE ROLE OF THE JOINT CONTROL CENTER

In the type of urban operations we emphasize in this report, we think it is unlikely that sensor detection of weapons, adversary personnel, or vehicles will lead to lethal fires being put automatically on the target. Rather, we expect there to be at least one human servicemember in the loop between the sensor and the shooter: a controller in the air or on the ground in the rear who has responsibility for a sector of the city in major operations in large cities or for the entire city in smaller operations. Controllers would develop situational awareness from ground-sensor inputs, communications with supported ground forces, airborne imagery, and background intelligence on expected adversary operations. They also would have knowledge of major civilian and friendly military activities planned for that day in their sector. For these concepts to work, the controller must have access to a 3-D database, be supported by sophisticated software that aids decisions by providing basic information about target coordinates (e.g., what building, what floor, who is thought to be there normally, where known friendly and adversary forces are), and have the power to authorize lethal fires. We illustrate this process in the following paragraphs.

Imagine a ground sensor detecting weapons moving through a building entrance, alleyway, or some other constricted feature. The controller's console gives an alert with basic information about the situation. For example, the standard message might say something like "Alert: Weapons, Type: Long-barrel small arms, Count: five and counting, Location: Lat, Long, altitude." The controller would select a database check (or perhaps this would be triggered simultaneously with the alert), and the coordinates would be compared to the urban database, providing additional information (Location: alley between Palms apartment building, 2100 East St, and abandoned warehouse at 2200 East St, Adversary forces: No current reports, Friendly forces: Foot patrol 5 blocks to east moving toward location, Civilians: Apartment building occupied, Recent operations: Friendly patrol ambushed 2 blocks west on 8/17/05).

Most ground systems would have multiple sensors to reduce false alarms and allow target ID. In this case, we envision a low-light TV or IR camera on the same system that detected the weapons. Alternatively, the camera might be located on a different system or airborne. The camera would be turned on and slewed automatically to the location where the weapons were detected. In this case, made simple for the sake of the illustration, the controller is able to observe the suspect personnel setting up an ambush in the alley. With many options at this point, the controller can request additional ground forces to surround and attack the ambushers or can bring in airborne fire support. In this case, the controller alerts the friendly patrol and also an AC-130 or other fire-support platform orbiting over the city. The coordinates are uploaded to the AC-130, and a glide bomb is dropped on the adversary forces.

Precisely because of the complexity of these situations and the need for superior judgment, we see the controller playing a critical role in integrating airborne surveillance and fire support assets in urban operations. Many difficult command and control issues would have to be resolved before this system could be put in place. Such a control center would be a cross between a standard ground-element command center and a Combined Air Operations Center. Given the level of integration required, a joint operations center for urban operations would need to be staffed by airmen, soldiers, sailors, and marines to ensure that all the necessary expertise is available.

In this chapter, we presented new concepts to accomplish some of the more important and vexing operational tasks confronting U.S. joint forces in urban settings. We focused on capabilities that could plausibly be fielded within a decade rather than on long-term possibilities. Nevertheless, the capabilities envisioned here will not just happen; to be realized, they will require focused R&D and prototype development. The next chapter identifies and assesses the state of the art in six technology areas that have promise for improving the contribution of aerospace forces to joint urban operations.

Chapter Six

ENABLING TECHNOLOGIES FOR
URBAN AEROSPACE OPERATIONS

INTRODUCTION

Aerospace forces have made important contributions to urban operations from World War II to the present (see Appendix D). Chapter Three identified some important limitations on the use of these forces in situations with very strict ROE and tight political/legal constraints. Although many would not even consider the use of aerospace forces for some of these operations, Chapter Five illustrated many ways that such forces could accomplish tasks that today can only be done by putting ground forces into very risky situations. With the right investments, DoD and the USAF can develop new capabilities that will allow the United States to achieve key objectives in urban operations more efficiently, minimizing risks to friendly forces in the process.

Six technology areas have promise for improving the contribution of aerospace forces to urban operations:

- Three-dimensional modeling of urban environments

- Communication and navigation systems

- Sensor technologies

- Sensor fusion

- Air-launched sensor platforms

- Limited-effects munitions.

In this chapter, we discuss both the latest advances in these areas and the technical hurdles that must be overcome to make the capabilities explored in Chapter Five a reality.

THREE-DIMENSIONAL MODELING OF THE URBAN ENVIRONMENT

Key Functions and the Scope of Air Force Involvement

Building high-quality 3-D maps and effectively integrating them with geospatial information represent a serious challenge for the Air Force, but one with a potentially great consequence for both air and ground operations in urban areas. The challenge can be divided into four key elements:

- Acquiring and processing data—acquiring current 3-D maps of urban areas and producing raw Digital Elevation Models (DEMs) of areas of interest

- Extracting terrain and surface object features for classification

- Associating feature data with geospatial information

- Updating the scene with dynamic information.

Air Force decisionmakers must tailor their technological investments in proportion to the extent and nature of their participation in each of these four key elements. The USAF needs to determine which, if any, of these elements are most appropriate for investment, and also how to address the division of responsibilities with organizations such as the National Imagery and Mapping Agency (NIMA). Of particular concern is the extent to which it is appropriate and necessary for the Air Force to undertake these activities independently rather than rely on other agencies for critical products. These issues, while not strictly technological, are central to understanding the relevance of different technology options.

The first phase of the urban modeling effort is building the baseline map for the area of operations, a task usually done and updated by NIMA in the course of normal tasking requirements. A Digital

Elevation Model consists of sets of latitudes and longitudes, along with height information relative to a standard geodetic datum.[1] High-fidelity DEMs have postings every few meters.[2] A typical DEM will have a portion reflecting the estimated contour of the underlying terrain, and a second portion containing a model of features, both natural and man-made, that overlie the terrain. Because the data acquisition-phase requires specialized collection platforms and sensors, it usually attracts the most attention.

A geospatial information system (GIS)[3] can be used to store the DEMs, along with other data of interest, and thereby greatly increase the value of what is collected by specialized assets. For instance, having a DEM with just the observed elevation (terrain with features) is useful for many applications such as route planning for low-altitude aircraft and missiles. However, when the data are married to other geospatial information (e.g., whether the object on top of the terrain is a stand of trees or a building, the object's function, street address, type of construction), the value of the data increases dramatically, along with the number of potential users.

The value of the data is increased further when dynamic information is added to the static data in the GIS, providing a current situational picture. This addition involves sensing dynamic and emerging static elements (obstructions in roadways, damaged buildings, etc.) as in the first stage, updating DEMs with information on feature changes and alterations of objects on the surface, then inserting transitory elements such as personnel and vehicles.

Most of the activities outlined above are classic intelligence-collection functions. Building the initial DEMs, compiling basic information on the area of operations, and populating the GIS fall within the domain of the intelligence community, with some division of responsibility between national and theater components.

---

[1] For a good introduction to mapping issues, see Defense Mapping Agency (now NIMA) *Geodesy for the Layman*, DMA TR 80-083, which is available at ftp://ftp.nima.mil/pub/gg/geo4layman/Geo4lay.pdf

[2] A *posting* is an average of elevation readings within a given area.

[3] A *GIS* comprises maps and data associated with particular locations. For example, in a real estate GIS, clicking on a home icon (on a map screen) may bring up the number of rooms, square footage, date built, assessed value, and other information.

Obtaining updates on features in the DEMs, adding dynamic elements, and ensuring the connection with both weapon systems and command and control elements is the business of the services. Capabilities overlap, particularly in building DEMs. If feature changes over a significantly broad area are to be included for purposes of mission planning and weapon employment, then many of the data-collection and -processing elements necessary for basic DEM construction could either be under the control of the service or levied on the national intelligence and mapping agencies as additional requirements for timely updates. Failure to clearly spell out responsibilities and funding obligations increases the likelihood that gaps will develop. Moreover, some technologies, if properly deployed, can be used to bridge potential gaps in capabilities.

In the next six sections, we describe three techniques for collecting data to construct 3-D urban models—laser radar mapping, stereo-scopic electro-optical imaging, and interferometric synthetic aperture radar (InSAR); compare the three techniques; discuss trade-offs involved in selecting collection platforms; and describe the computer software the warfighter needs to display and manipulate the models.

Laser Radar Mapping

Used for decades in the civilian sector for conducting airborne surveys, laser radars currently dominate the imaging field. Laser radars are similar to conventional pulsed radars, except that light pulses are emitted instead of radio-frequency pulses. Typical commercial laser radar systems, including ancillary equipment, weigh between 100 and 250 lb.[4,5] The laser is mounted pointing downward from a stabilized platform on the aircraft. The aircraft position is established to within a fraction of a meter, using differential GPS.[6] The laser transmitters are usually solid-state neodymium yttrium aluminum

---

[4]D. Henderson, "Lidar System Finds Fault with Trees," *Photonics Spectra*, August 1999, pp. 22–24.

[5]Optech Inc., "Airborne Laser Terrain Mapper," available at http://home.ica.net/~esk/altm.html (downloaded in September 1999).

[6]Differential GPS is a way of increasing the accuracy of the GPS signal by transmitting a signal containing correction information derived from a receiver at a surveyed location on the ground.

garnet (Nd:YAG) devices operating in the band just above 1 µm, which lies just outside the visible band, in the IR. At this wavelength, there are minimum-altitude guidelines[7] to avoid damaging eyesight on the ground; techniques are being introduced into military systems to shift the wavelength to the eye-safe band near 1.6 µm. In principle, other forms of aided GPS could be used, such as the pseudolites discussed later in this chapter.

The narrow laser beam scans a swath below the aircraft, precisely measuring the range to the ground and the angle of deflection of the beam as it sweeps to the left and right of the aircraft's nadir spot.[8] In some systems, multiple beams are transmitted simultaneously to increase the swath width. Assuming typical commercial parameters— e.g., a swath angle of 20°, aircraft altitude of 1 km, and speed of 250 kt, we obtain a ground swath 730 m wide and an area sweep rate of 330 km^2 per hour. The aircraft will traverse many parallel swaths if a large area is to be covered. At the stated rate, the city of Los Angeles could be mapped in 3 to 4 hours. The range and angle data are recorded digitally and are combined with the aircraft's position record to yield a digital elevation map of the terrain. Typical elevation accuracy is 0.05–0.15 m.[9] (See Figure 6.1 for a sample image produced by laser radar mapping.)

Ideally, archival urban models could be compiled during peacetime and accessed as needed for military purposes. However, updates during wartime or other emergencies may be required, because urban areas are in constant flux and the programmed refresh rate of the database will not always maintain adequate currency. In light of the potential urban threats discussed earlier in this report, it would be desirable to collect data from high-altitude UAVs.

---

[7]Laser radars are routinely used by private-sector mapping companies to produce 3-D maps of cities. The radar-equipped aircraft must fly above certain minimum altitudes to avoid the risk that *anyone* on the ground might have their eyes exposed to dangerous levels/frequencies of laser energy.

[8]N. Savage, "Lidar Sensor Sees Forest and the Trees," *Laser Focus World,* May 1999, pp. 71–72.

[9] This level of accuracy is achieved by processing out the GPS bias.

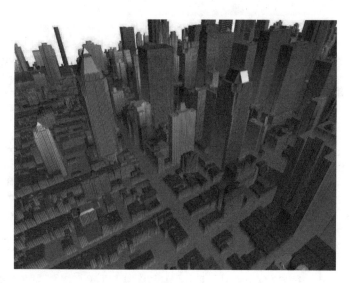

SOURCE: Intermap Technologies Inc.
NOTE: Height information in this laser map has an error of 0.3 m.

Figure 6.1—High-Resolution Laser Radar–Derived Digital Elevation Model of San Francisco

The ability to collect from high altitude is perhaps best illustrated by the Mars Orbiter Laser Altimeter (MOLA), carried aboard NASA's Mars Global Surveyor. In March and April 1999, MOLA collected 27 million laser measurements of the Martian surface through the thin Martian atmosphere, to form a 3-D map of much of the planet. To operate at the orbital altitude of 400 km, the laser receiver was equipped with a large, 50-cm-diameter, parabolic mirror and a sensitive silicon-avalanche photodiode detector.[10,11]

---

[10]K. Lewotsky, "Mars Surveyor Altimeter Flies High in Orbital Test," *Laser Focus World*, November 1997, pp. 43-46.

[11]Silicon photodiodes are one type of semiconductor device used as an element in CCD arrays, which are commonly used in focal planes for commercial digital cameras. Avalanche photodiodes are even more sensitive than the photodiodes ordinarily used in cameras. They have some of the properties of image intensifiers, such as the ability to detect smaller numbers of photons by amplifying the electrical signals produced by the photons impinging on the detector.

The main impediments to high-altitude operation are weather and degraded angular resolution, which results in degraded horizontal resolution for the map. If the MOLA laser, with its beam divergence of 0.46 milliradians (mr),[12] were used for mapping from Global Hawk at 60,000 ft, the spot size on the ground would be 8.5 m.[13] Measurement accuracy, e.g., for obtaining GPS coordinates of a building's corners, can be improved if the laser samples are spaced only a fraction of the spot size apart. An improvement factor of 2 to 3 seems attainable, which would be adequate for the more-demanding applications of guiding weapons up to the terminal phase.

Stereoscopic Electro-Optical Imaging

Stereoscopic imaging is the basis for human depth perception. When we view a nearby object, the observation angle from each of our two eyes is slightly different, an effect known as *parallax*. The amount of parallax displacement depends on the range to the object and the separation between our eyes. Without our having to think consciously about it, our brains are able to convert the parallax offset into an estimate of range, which we perceive as depth. But, how are the two "eye views" seamlessly combined into one? Conceptually, the eyeviews are warped until every point is remapped into a single, unified picture. The amount of warping required to match up a point is a measure of its parallax. The tricky part is that identifying corresponding points sometimes requires paying attention to the *content* of the image, which is easy for brains but hard for computers.

An image collected by a single spaceborne electro-optical (EO) sensor contains only two-dimensional information, analogous to a single human eye. It is possible to recover the third dimension by measuring parallax between images obtained from two satellite or aircraft locations, but the positions and pointing directions of the optics must be known precisely.

---

[12]A *milliradian*, abbreviated mr, is approximately 0.0573°. At a range of 1 km, an angle of 1 mr subtends exactly 1 m, i.e., 1/1000th of the range—a convenient unit.

[13] The smaller the spot size, the higher the resolution.

Intensive research is under way to completely automate the process of extracting depth from stereo imagery. Currently, however, smart workstations and software tools exercised under human supervision are relied on for such extraction. The near-real-time production of urban models that laser radars can offer today may not be possible with EO for some time.

Nonetheless, EO can provide higher-resolution information on exterior details of buildings, e.g., the placement of windows, thickness of walls, and construction materials, than can laser radar. Typical large EO sensors used for reconnaissance have ground resolution of 0.6 to 0.3 m from the Global Hawk altitude of 60,000 ft. Spaceborne EO sensors could have ground resolution as good or better than this from a 600-km orbit.

Interferometric Synthetic Aperture Radar

Interferometric synthetic aperture radar (InSAR) is a technique for coherently combining two SAR images taken from two slightly offset positions, either simultaneously with two or more antennas, or separated in time with one antenna (see Figure 6.2).[14] The coherence property refers to preserving the phase information in the image, which stems from the wavelike nature of electromagnetic signals. When SAR images are first synthesized, each resolution cell has associated with it an amplitude and a phase angle between 0 and 360°; when displayed, the phase information is usually suppressed. The phase is a measure, but an ambiguous measure, of the two-way distance, d, between the antenna and the point on the ground.[15]

Consider two SAR images of the ground obtained with antennas displaced along the cross-track direction from one another. The two-

---

[14]H. Zebker et al., "Topographic Mapping from Interferometric Synthetic Aperture Radar Observations," *Journal of Geophysical Research,* Vol. 91, April 1986, pp. 4993–4999.

[15]If the distance is a perfect multiple of the wavelength, the phase is zero. If the distance is N wavelengths plus a half-wavelength, the phase is 180°, i.e., halfway between 0 and 360°. In general, if the distance is N wavelengths plus a fraction, F, of the wavelength, the phase is 360° × F. Phase is spoken of as being related to the distance, *modulo the wavelength*.

RAND *MR1187-6.2*

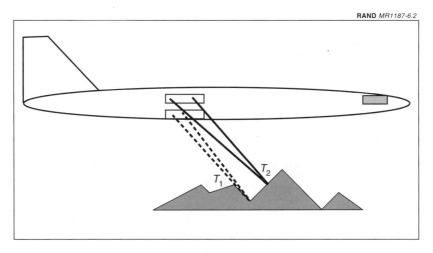

Figure 6.2—Single-Pass InSAR Geometry

way distance T_2 from the top antenna to some resolution cell on the ground is slightly longer, by an amount Δd, than the two-way distance to the bottom antenna, T_1. If we subtract the phases between the two SAR images, producing a *phase-difference map*, this resolution cell will have the phase corresponding to Δd.

If, instead of the ground, the resolution cell in question contains a raised structure, the difference in distance from the two antennas to the roof of the structure will not be Δd but some other value, $\Delta d'$, depending on the height of the structure. In this way, the phase-difference map encodes information about a third dimension—elevation—that is not available in a single SAR image (see Figure 6.3).[16]

---

[16]There is a slight hitch, however. The phase is related to Δd ambiguously: Adding (or subtracting) integer multiples of the wavelength to (from) Δd yields the same value of phase. But, to compute the height, precisely *how many wavelengths* must be known. This number can be determined by looking at the trend of phases from resolution cell to resolution cell. For example, if the sequence of phases 10°, 120°, 260°, 30°, 100°, 190°, 320°, 60°, 220° is seen along a string of neighboring cells, one infers that between 260° and 30°, and between 320° and 60°, the phase passes through 360° and Δd has increased by a wavelength. This scheme works as long as the height differences between adjoining cells are not so large that Δd changes by more than a wavelength. For this reason, skyscraper-studded urban canyons are not ideal venues for InSAR mapping.

SOURCE:Intermap Technologies Inc.
NOTE: This image consists of a 2.5-m-resolution orthorectified image draped over a 3-m-vertical-accuracy DEM.

Figure 6.3—InSAR Combined with Panchromatic Image Overlay of Howard Air Force Base, Panama

SARs have difficulties in urban settings when buildings are high or if the streets are narrow. As noted in Chapter Four, imaging this geometry requires a steep depression angle to look down to street level. In this geometry, SARs suffer from the *nadir hole* problem. To understand the origin of the nadir hole, it is necessary to explain how SARs form an image. SARs obtain slant-range information (see Figure 4.3) in the same way some conventional radars do, using the time delay and amplitude of reflected pulses. Phase or frequency coding causes the SAR pulses to have large bandwidths. When the pulses are re-

---

Recent techniques introduced to cope better with effects of phase ambiguity include using multiple frequencies or multiple baselines, the latter typically involving three antennas instead of two. See G. Corsini et al., "Simulated Analysis and Optimization of a Three-Antenna Airborne InSAR System for Topographic Mapping," *IEEE Transactions on Geoscience and Remote Sensing*, Vol. 37, No. 5, September 1999, pp. 2518–2529.

ceived and processed, the result is very fine slant-range resolution.[17] In composing a SAR map, a simple transformation is performed to convert the slant-range data into ground range. If the SAR is viewing the scene at low depression angles, slant range and ground range directions are almost the same, and the resolution is not seriously degraded in transforming between them. At higher depression angles, the resolution along the ground is degraded in proportion to the secant of the depression angle; for example, a 1-m-resolution SAR has a ground-range resolution of 2 m at 60°, 2.9 m at 70°, and 5.8 m at 80° depression angle. At sufficiently high depression angles, the ground-range resolution is so poor that the imagery is not worth collecting—hence, the nadir hole.

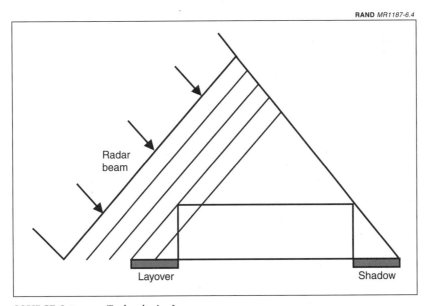

SOURCE: Intermap Technologies Inc.

Figure 6.4—Layover and Shadowing Effects in SAR Imagery

---

[17]Azimuth information is obtained by Doppler-processing a sequence of these pulses as the sensor moves along its flight path or orbit.

Urban areas can also be troublesome to SARs because of shadowing and layover effects (see Figure 6.4). *Shadows* occur where the signal is obstructed, producing regions in which all information is lost. Shadowing can be mitigated by combining images from different angles. *Layovers* occur because SARs cannot distinguish between objects that are taller and objects that are merely closer—i.e., the tops of buildings appear to be lying on the ground in the foreground of the structure. Layover effects can be removed by InSAR processing, which restores the height information.

In their finest mode, typical military airborne SARs have resolution between 0.3 and 3 m. The SARs for both Predator and Global Hawk have 0.3-m resolution. As noted in a preceding paragraph, these values increase in the ground-range dimension at high depression angles.

SAR payloads for both aircraft and satellites have been used to generate InSAR maps. Although multipass InSAR imaging does not require additional hardware (each of the two platforms uses a single antenna), single-pass InSAR requires that additional apertures be offset along the cross-track direction. The primary requirement for handling InSAR data is that the coherent images (with the phase data intact) be datalinked to the processing center.

A good deal of work has been done in the last few years to automate the process of extracting height from InSAR maps (counting the phase cycles through 360° a process called *phase unwrapping*) and creating 3-D databases for computerized display. Recent improvements in the algorithms have reduced requirements for human intervention to a low level. However, the problems associated with building extremely precise maps in urban areas still require a great deal of human intervention.

Both civilian and military agencies have plans for spaceborne collection of worldwide topographic data using InSAR mapping. This strong interest is driven primarily by the prospect of providing high-resolution digital terrain elevation data (DTED). However, another particularly useful side benefit with InSAR maps is that they can detect change exquisitely, at the level of inches of vertical deflection. In cities, new construction shows up readily as a random mismatch of the phases between the "before" and "after" InSAR maps. Such change detection could also be used to observe changes in roads that might indicate obstacles or other key features changing within a tactically relevant time-scale.

Comparison of 3-D Imaging Technology

Each of the three methods for obtaining 3-D urban data discussed has its pros and cons with respect to weather limitations, automated processing, resolution, and constraints on slope or depression angle.

Only the InSAR has all-weather capability. Clouds or fog can obscure the ground and deny useful imagery to laser radar and EO sensors, which operate in the optical bands. It is therefore advisable to use laser radar when possible, but have InSAR as a backup, when developing militarized payloads.

Each sensor technology offers other features that are useful and unique. Laser radar provides the most automated and rapid processing. EO provides the best horizontal resolution, and a great deal of qualitative information about structures. InSAR processing is reasonably well-automated, provides the best vertical resolution, and is all-weather, but it is somewhat constrained in imaging steep slopes and high depression angles.

It is likely that high-quality urban mapping will depend on combining data from all three techniques in the future. Today, a number of agencies routinely produce maps combining pairs of sensor types, most frequently EO and SAR, or EO in several bands. Although this combining is usually still performed by a human operating a workstation, the operator is using tools that are rapidly increasing in sophistication.

As to the likelihood that fusion and accurate geolocation of multisensor images will be fully automated before too long, we adduce three factors for optimism:

- No underlying physical barriers to broach or looming engineering paradigms to shift

- The presence of a large and active community, including universities, civilian agencies such as NASA, and the intelligence community, devoting significant resources to the problem

- Expansion of processing power at the rate predicted by Moore's Law (doubling every 18 months) for at least another decade, as expected by computer engineers.

Platform Trade-Offs

The platforms used for data acquisition fall into two categories: air-breathing and spaceborne. The factors governing the choice of platform are the sensor type, area of coverage, level of threat, and political sensitivities.

The types of air-breathing platforms that can be used for imagery collection vary considerably, including small UAVs, small manned aircraft, business jets, fighter aircraft, high-altitude long-endurance UAVs, high-altitude manned aircraft, and larger transport-type aircraft.

Microwave and laser radars and electro-optical sensors have been deployed on both aircraft and spacecraft; in aircraft, they have been deployed over the full range of altitude. As noted in Chapter Four, the presence of manportable air defenses and light anti-aircraft artillery can preclude the use of UAVs like Predator at low altitude. Radar-guided tactical SAMs can be overflown with high-altitude aircraft such as U-2s and Global Hawks. Modern high-end SAMs such as SA-10s and SA-12s can engage these aircraft, thus potentially denying access to all but spaceborne collectors before a successful SEAD campaign.

Lasers and EO cannot penetrate cloud cover and dense fog and are degraded by atmospheric attenuation. Consequently, subject to threat considerations low-altitude operation has the advantage of extending the spectrum of weather conditions under which optical sensors can collect data. Low-altitude operation is also favored for optical sensors because their resolution improves with decreasing altitude. Optical sensors can operate at nadir; SARs cannot, because they depend for image formation on slant-range resolution, which degrades at nadir. The area sweep rate for EO depends primarily on the sensitivity, efficiency, and size of the detector elements, the size of the aperture and focal plane, and the speed of the aircraft. Since EO sensors are passive and not usually power-limited, they can generally be made to cover area more rapidly at high resolution than can SARs.

Microwave radar is nearly impervious to weather. However, heavy rainfall will blind it at frequencies above 10 GHz. The resolution of synthetic aperture radar does not depend on altitude as long as there is sufficient power to preserve a high signal-to-noise ratio. The demands for higher power and processing throughput usually decrease the area coverage rate with finer resolution.

Spacecraft offer the advantages of worldwide coverage, better covertness, and the ability to operate over areas denied by threats or political considerations. Disadvantages include higher initial procurement costs, greater complexity, long revisit time (depending on orbit), and unfavorable absentee ratio. *Absentee ratio* refers to the long time the spacecraft spends outside the viewing range of the area of interest due to its orbital motion. Many surveillance instruments use change detection, which requires a baseline image against which to compare the new image. Spaceborne systems have a unique ability to obtain baseline data without requiring permission of the host country for imaging.

Software Exploitation of 3-D Urban Maps

In the commercial sector, Digital Elevation Models (terrain plus features above the terrain) are extensively used to support a wide range of activities, from helping urban planners deal with land-use decisions and make water-drainage assessments, to predicting fires and exploiting telecommunications. Indeed, the wireless telecom munications industry's use of DEMs for siting cell-phone towers represents one of the more prominent applications of this kind of technology, and there are many parallels with applications in the military sector.

DEMs provide obstruction data, along with other parameters associated with signal propagation, which enable the optimal placement of transmitters. Analogous applications for the military include the placement of terrestrial communications relays, observation platforms, or anything that requires line-of-sight calculations, such as countersniper operations.

In the military context, software for exploiting the vast amount of data coming from capable sensors has received somewhat less attention than have sensors and surveillance and reconnaissance platforms. The problem of fusing and presenting data to decision-makers is, in many respects, a more technically challenging problem than collection. For instance, ambiguities in target identification, geolocation, and sensor measurements make associating multi-source information a major challenge, as do difficulties in constructing the underlying GIS. Thus far, the complexity involved in designing generalized fusion techniques has rendered the building of an effective, full-scale fusion system infeasible.

For achieving fusion, there are circumstances under which simpler approaches will yield substantial results. For example, simply combining essentially raw data from sensors with information stored in a GIS is often adequate to allow a skilled human operator to understand what is occurring.

The first step toward building a useful product is to combine imagery and DEMs to generate 3-D perspectives of target areas. These perspectives can be viewed using computer-based visualization, a technique employed for some time in mission-planning systems. The next step is to add in geospatial data and apply predictive algorithms. A common GIS can be used for the basic infrastructure, which avoids the cost of designing a system from scratch while providing the ability to capitalize on a wealth of commercially derived software. However, the most difficult parts of the problem remain: populating the database with current information and interfacing with a host of real-time intelligence and battle-management systems.

Once the database is formed, there are many pathways toward military utility. Some, like the countersniper example discussed in Appendix C, exemplify the use of very basic information to assist in an operation. More-sophisticated applications might allow the operator to formulate queries to understand the construction, use, or other aspects of structures in an area of interest.

COMMUNICATIONS AND NAVIGATION TECHNOLOGY FOR THE URBAN ENVIRONMENT

UAV Relays

Communication and the reception of navigation signals within the urban environment are bedeviled by multipath[18] and obstruction, problems that are further compounded by co-channel interference and Doppler shifts when receiving data from a network of mini- or micro-UAVs. Jamming is also a potential problem, which could degrade control signals to the netted UAVs, or data uplinked to a satellite or relay UAV. Finally, signals from implanted sensors might be intercepted and geolocated, leading to the unit's destruction or compromise.[19]

Solving these problems calls for signaling techniques having inherent antijam and low-probability-of-intercept characteristics, such as spread spectrum or impulse radio. Both approaches allow many channels to coexist in the same band and can largely eliminate multipath effects through appropriate processing.

The weakening of signals by obstruction can be exacerbated by low power if the signal source is a MAV or propagation loss if the signal source is an in-building radio. MAVs, weighing a fraction of a pound in total, must restrict their communications payloads to a weight of several grams, with power consumption of perhaps 1 W. Obstruction from using cell phones for communications from inside buildings may reduce effective power to a fraction of a watt in passing through concrete block walls.

---

[18]In topologically complex environments, such as urban areas, radio signals from a transmitter arrive at the receiver by diffracting and reflecting from the ground and from buildings. Since the signals arrive by way of a number of radio paths, the combined signal is called a *multipath signal*. The relative delays between the component signals can cause them to mutually interfere. This interference can reduce the resultant signal power or can produce overlap between adjacent symbols in the data stream—in either instance, possibly leading to errors in the received signal.

[19]P. Johnson et al., "Micro Aerial Vehicle Communications Architecture for Urban Operations," *AUVSI '98*, Association for Unmanned Vehicle Systems International (AUVSI), 1998.

The common view that commercial satellites are the panacea for these problems wavers under scrutiny. Recent problems in operating Iridium phones in the city and the very real possibility of uplink jamming should raise concern about the robustness of commercial satellites. Communications relays on high-altitude UAVs such as Global Hawk can offer line-of-sight, or near line-of-sight, links with reachback to fusion centers via additional UAVs, or links to commercial satellites that are over the horizon from ground-based jammers.

It is important to recognize that data links from MAVs and implants in the city are very challenging even when UAV relays are present. Even with image compression, a data rate on the order of 100 kilobytes per second (kB/sec) may be required to transmit imagery. To transmit within an extremely tight weight and power budget will require the development of specialized receivers-on-a-chip. The impulse radio techniques discussed in the next subsection may also contribute to the solution.

DARPA has an ongoing program to develop a UAV relay payload for Global Hawk, called Airborne Communication Node (ACN).[20] The concept involves servicing a plethora of in-theater relay needs, including broadcast to forces on the move, theater paging, handheld radio, Joint Tactical Information Distribution System (JTIDS) support, position location (PLRS, EPRLRS), and acting as gateway among dissimilar radios, e.g., SINCGARS, HAVE QUICK. Electromagnetic self-interference and interoperability issues hamper meeting all these needs within a single payload. Software radios that can bridge different modulations and protocols are being developed to address some of these problems. Carefully tuning payloads to specific missions may be the key to success for this very important capability.

Through-the-Wall Communications

To potentially overcome many of the obstacles to communications in the urban environment, communications equipment using ultrawideband (UWB) impulse waveforms is being developed and tested.

---

[20]DARPA, "Airborne Communications Node," briefing, Washington, D.C.: DARPA/Sensor Technology Office (STO).

Promising features of this technology include low probability of intercept (LPI), non-interference with nearby users, relative immunity to multipath and jamming, and ability to penetrate structures.[21]

Conventional communications is based on encoding data onto signals by modulating a carrier wave. Impulse radio is carrierless: Instead, millions of single-cycle nanosecond-duration impulses are transmitted per second, with randomized (but known to the receiver) interpulse intervals to make the waveform appear noiselike. The information content is encoded onto the pulse stream by pulse-position modulation (PPM): Binary 0 is delayed, and binary 1 is advanced a fraction of a nanosecond relative to the nominal pulse position. This scheme is referred to as *time-modulated UWB*.[22]

Each data symbol consists of a stream of zeroes and ones, 10^2 to 10^3 bits long. The symbol is detected by coherently summing the energy from this set of bits. The individual bits can have such meager energy that they are below the level of ambient noise; yet, the processing gain achieved by coherent summation allows the receiver (with knowledge of the time code) to detect the symbols.

The ability to hide the time-modulated waveform in noise makes it difficult for hostile receivers to intercept. The receiver is resistant to jamming because it is receptive to signals only when they come within the short time interval during which a pulse is expected and because it is designed to sense the rapid increase in signal amplitude at the start of a pulse. The waveform is also resistant to multipath, the potentially destructive interference between signals arriving along slightly different paths, for example, a direct path through the air and one that bounces once off the ground or a building. The very short pulse width ensures that only propagation paths with lengths differing by as little as 0.3 m will mutually interfere—a situation that

---

[21]W. Scott, "UWB Technologies Show Potential for High-Speed, Covert Communications," *Aviation Week & Space Technology*, June 4, 1990, pp. 40–44; P. Withington, "Impulse Radio Overview," available at www.time-domain.com.

[22]"Time Modulated–Ultra Wideband Radio Measurement and Spectrum Management Issues," available at www.time-domain.com (downloaded September 1999).

applies to only a very small fraction of multipath bounces, even when communicating inside a building.[23]

A single user's transmitter is "on" only between 0.1 and 1 percent of the time. By employing different time codes, which are pseudorandom and have low cross-correlation, many users can operate on a non-interfering basis.[24] The few pulses that randomly invade a neighbor's time code will be overwhelmed by the coherent processing gain of the receiver using the correct code.

The typical center frequencies of impulse radio fall between 650 MHz and 5 GHz. Frequencies at the lower end can penetrate structures, such as concrete block walls, with minimal losses.

Impulse radios developed by Time Domain Corporation and Multispectral Solutions, Inc. (MSSI) are small and lightweight and have low power consumption.[25] Tests of MSSI's 1-W impulse packet radio, which weighs 1.9 kilograms (kg) and operates at a 9600-baud data rate, demonstrated the ability to operate a line-of-sight link successfully at a range of 20 mi. A handheld voice/data impulse radio developed by MSSI, weighing 1.1 kg, transmits data at a rate of 128 kB/sec, although, with its existing antenna, at rather short range. On the whole, this technology seems capable of supporting a variety of secure data links to high-altitude UAV relays from Special Forces inside buildings and from implanted sensors, mini-UAVs, and MAVs.

However, the future of UWB radios is threatened by concerns about interference with navigation systems.[26] The Federal Aviation Administration (FAA) is petitioning the Federal Communications Commission (FCC) to ban UWB radios on the grounds of potential interference with avionics and navigation units. Although the wide bandwidth of UWB does impinge on sensitive frequencies, impulse

---

[23]P. Withington, "In-Building Propagation of Ultra-Wideband RF Signals," available at www.time-domain.com (downloaded September 1999).

[24]R. Scholtz, "Multiple Access with Time-Hopping Impulse Modulation," *Proceedings of IEEE MILCOM '93*, Boston, Mass., October 11–14, 1993.

[25]R. Fontana et al., "An Ultra Wideband Communications Link for Unmanned Vehicle Applications," available at www.his.com/~mssi (downloaded September 1999).

[26]W. Scott, "UWB Industry Fate May Hinge on Review," *Aviation Week & Space Technology*, December 14, 1998, pp. 63–64.

radio's low power spectral density should allow room for some level of use. If UWB is shut down domestically, the loss of dual-use efficiencies will have cost ramifications for military-only systems.

Pseudolites

Pseudolites are ground-based or airborne transmitters that supplement or replace GPS for navigational purposes. The civilian world is interested in pseudolites primarily for improving accuracy. Civilian applications include precision approach and landing of aircraft and land surveying. Recent flight tests have been conducted to determine the value of integrating a pseudolite into the FAA's ground-based Local Area Augmentation System (LAAS). The LAAS is intended to bolster GPS to permit all-weather landings at airports.[27]

Military interest is centered on exploiting the pseudolite's shorter range to the user and possibly higher power in order to strengthen the GPS signal against jammers. In the urban environment, the pseudolite would function to counter jamming, propagation loss in the urban canyons, and multipath. Overcoming severe propagation losses in the city calls for an airborne pseudolite, which generally has a more direct line of sight to users in an area than does a ground-based pseudolite. Preferably, the pseudolite would be mounted on a high-altitude, enduring platform such as Global Hawk.

Several technological challenges have been encountered in developing airborne pseudolites:[28]

- Accurately determining the location of the pseudolite platform

- Transmitting ranging signals that can be received by GPS receivers

- Injecting pseudolite position data into a format compatible with existing GPS equipment

---

[27]L. Dorr, News Release from Federal Aviation Administration Technical Center, August 13, 1999. Available at www.faa.gov/apa/pr/.

[28]R. Greenspan et al., *Robust Navigation Panel Final Report*, Cambridge, Mass.: Draper Laboratory, Report CSDL-R-2833, 1998.

• Avoiding signal degradation from interference with satellite GPS signals.

None of these challenges appears overly daunting.

The feasibility of a pseudolite payload for Global Hawk is being investigated by DARPA/Sensor Technology Office (STO), and DoD is planning a UAV flight test in the near future. The concept is for a pseudolite with a sophisticated antenna to receive the GPS signal (for its own localization) by placing deep nulls on jammers. The pseudolite then broadcasts its own ranging signal, which can be picked up by slightly modified GPS receivers.[29]

IMAGING SENSOR TECHNOLOGY FOR URBAN OPERATIONS

A revolution in imaging sensor technology is under way and will profoundly affect the design of surveillance payloads in the next decade.[30] The advances are appearing in several domains, the most important being

• Large focal plane arrays (FPAs), in the megapixel class, in all the optical bands

• Large, uncooled FPAs operating in the IR bands

• Microsensors suitable for expendable implants and MAVs.

As digital cameras, large FPAs operating in the visible band have been highly commercialized, which has significantly reduced their cost. Large FPAs offer high resolution, large field-of-view, rapid readout, good dynamic range, and frame rates adequate to surveil scenes that change quickly.

Large cooled FPAs operating in the IR will have improved sensitivity, enabling them to detect in multiple bands for improved discrimina-

---

[29]"Pseudolites—A GPS Jamming Countermeasure?" *Flight International,* July 28–August 3, 1999.

[30]S. Horn et al., "Third Generation Sensors," *Proceedings of the IRIS Specialty Group on Passive Sensors,* Vol. 1, 1999, pp. 403–415.

tion. Their tolerance for higher operating temperatures will reduce weight and power requirements.

Uncooled IR FPAs have extensive commercial applications—for security, police work, medical sensing, traffic control, etc.—and the competitive dual-use market is already whittling down the cost of FPAs of modest size. The obviation of requirements for cooling and temperature stabilization will decrease the complexity and expense of these systems.

Microsensors employing large FPAs in both the visible and IR bands represent a completely new category of imaging sensors. They will be sufficiently small and light weight to serve as payloads for MAVs or as expendable implanted sensors.

Apart from the visible-band cameras, numerous technical challenges must be met before these new sensors are available for urban surveillance. The uncooled FPAs need improved sensitivity, and the large megapixel arrays, for which the commercial market may be thin, must undergo innovations to reduce their cost. The micro-sensors share these two challenges, as well as the need to miniaturize electronics, reduce power requirements, compensate for temperature deviations in lieu of stabilization, and find commercial markets. The challenges for cooled FPAs are to decrease non-uniformity, shrink pixel size to accommodate multiband detectors, and raise FPA operating temperature to minimize cooling loads.

In the domain of urban operations, all the imaging technologies mentioned above have worthwhile applications. As stated in Chapter Four, a major challenge for aerospace operations is the lack of high-resolution sensors that can identify adversaries who are potentially mingled with civilians or friendly troops. Confidently identifying people is not a strong suit of IR sensors, which lack adequate resolution except at very close range. Even then, IR sensors do not supply the characteristic cues that humans rely on to "check people out" in the visible band.

The most appealing solution is to employ a visible sensor having good low-light-level capability, supplemented by active laser illumination at night or when looking through windows into darkened rooms.[31] Researchers at Lincoln Laboratory are developing a silicon charge-coupled-device (CCD)-based microsensor with sensitivity into the near-IR that fills this niche. By selecting a laser wavelength just beyond the visible band, the laser can be operated covertly from a mini-UAV, MAV, or, in some cases, a high-altitude UAV.[32]

An example of the weight trade-offs with range for a CCD sensor are shown in Figure 6.5, for daylight in clear weather. The CCD array has a dimension of 640 × 480 pixels, with detector spacing of 24 μm. The f-number of the optics is 2.7. In the figure, the optics diameter is scaled up as the weight increases, resulting in extended range per-

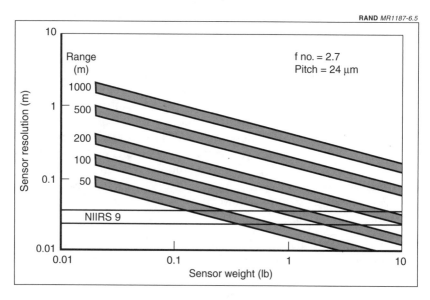

Figure 6.5–Imaging-Sensor Weight and Performance Trade-Offs

---

[31]M. Cantella, "Micro Air Vehicle Sensors," *Proceedings of the IRIS Specialty Group on Passive Sensors*, Vol. 1, 1999.

[32]The Electro-Optical Targeting Branch of the Air Force Research Laboratory has been developing a system for imaging tactical targets at ranges from 10 to 24 km, using a laser illuminator and a time-gated short-wave IR camera. See F. Kile et al., "Enhanced Recognition and Sensing LADAR (ERASER) Long Range 2-D Imaging," *Proceedings of the IRIS Specialty Group on Active Systems*, Vol. 1, 1998, pp. 19–31.

formance. We see that, for a mini-UAV of size intermediate between a Sender and Swallow and having a payload capacity of 5 lb, NIIRS-9 quality can be achieved at a range of approximately 150 m. A MAV with a 15-g payload achieves this NIIRS level at approximately 25 m.

A mini-UAV is likely to be detected acoustically, then optically, at a range of 150 m by an alert human "target." Therefore, it might be better utilized for surveillance tasks short of human identification. Although detectable at 25 m, the MAV can probably fly in under the cover of night and perch on buildings adjacent to the target building.

An added advantage of this approach is that the MAV need not exhaust its batteries and cease functioning in a mission of just 1 hr. The MAV also has an opportunity to recharge batteries, as discussed in Appendix B.

The resolution associated with NIIRS 9 depends on design details of the optics, focal plane, and processing, as well as on atmospheric and lighting conditions; however, for the clear-weather/daylight case represented in the figure, a typical NIIRS-9 resolution is in the range of 2.5 to 3.8 cm.[33]

A more detailed performance evaluation shows that a sensor designed to "identify" in daylight is reduced to "recognition" capability at twilight and must resort to laser illumination to retain recognition capability if lighting degrades further. If the optics is stationary, either housed in an implant or a parked MAV, resolution degrades more gracefully with low lighting, because integration time can be increased and more photons collected without motion-related smearing of pixels.

---

[33]J. Leachtenauer, "National Imagery Interpretability Rating Scales: Overview and Product Description," *ASPRS/ASCM Annual Convention and Exhibition Technical Papers: Remote Sensing and Photogrammetry*, Vol. 1, 1996, pp. 262–272; J. Leachtenauer et al., "General Image Quality Equation: GIQE," *Applied Optics*, Vol. 36, 1997, pp. 8322–8328.

NON-IMAGING SENSOR TECHNOLOGY
FOR URBAN OPERATIONS

This section discusses some non-image options for overcoming the difficulties associated with imaging sensors, such as bandwidth problems and weather restrictions, and the challenge of interpreting massive amounts of imagery.

Seismic and Acoustic Sensors

Seismic and acoustic devices have been considered an integral part of unattended ground sensors (UGS) for decades, during which steady advances have been made in sensor technology, particularly in signal processing.[34] Examples of already-developed UGS devices include the Remotely Monitored Hand-Emplaced Battlefield Sensor System (REMBASS), which adds infrared and magnetic sensing to acoustics; IREMBASS (the improved version of REMBASS); and Alliant Technology's SECURES, a commercial acoustic countersniper network. Further development of these devices in under way by DARPA and the services. In this subsection, we focus on vehicle surveillance using acoustic and seismic devices. For a discussion of countersniper sensors, refer to Appendix C.

The battlefield sensing problem involves detecting, classifying, and geolocating wheeled and tracked vehicles. Detection ranges of hundreds of meters to kilometers are typical, and networked sensors are able to establish locations with modest accuracy by measuring time-difference-of-arrival (TDOA) to several sensors in the network. Geolocation is complicated by the need to correctly associate detections made by different UGS in the network.[35] Ambiguities in association multiply as the density of vehicles increases, because the sensors have poor angular resolution, there are mismatches in the signals as a result of different Doppler shifts experienced at the nodes,

---

[34]"Integrated Acoustic Sensors for RFPI," *Proceeding of the 5th Battlefield Acoustics Symposium,* Ft. Meade, Md., September 23–25, 1997, pp. 326–357.

[35]*Association* refers to the matching up of detections or tracks of a target by the same sensor or different sensors, so that it is apparent that all the data refer to the same target. The likelihood of confusion increases if the sensors individually have poor resolution, which results in a group of targets being perceived as an undifferentiated blob.

and detection dropouts caused by the masking of distant sources by proximate sources leave too few independent detections to perform TDOA.

All the complexities of the battlefield are amplified in the city. Beyond short ranges, high levels of ambient traffic and anomalous propagation can make hopeless the resolution of ambiguities in associating targets. An appropriate role for seismic and acoustic sensing in the urban environment is to count and classify vehicles passing at close range over key roads or through key intersections.

Classification by seismic and acoustic sensors is based on recognizing the characteristic frequency content of a particular class of vehicles.[36] Acoustic and seismic frequency spectra consist mostly of lines, which typically form a series that are integer multiples of a fundamental frequency. These series, consisting of a fundamental and its harmonics, can be traced to specific physical phenomena. The engine's *acoustic signature* comes from the exhaust, and (in tracked vehicles) from the back cogwheel. The *seismic signature* (detected in the vibration of the ground) comes from the roller wheels moving over the track elements and, to some extent, from coupling into the ground of the acoustic sources.

There remain significant challenges in signal processing for classification. Data describing line series for a wide variety of vehicles have been collected by, for example, Sandia National Laboratories. Algorithms need to be perfected that can classify line series in the presence of line overlaps and that can deal with the overlap of whole spectra occurring when closely spaced vehicles pass by.

Through-the-Wall Radar

Two radar technologies have been developed in recent years that are referred to as "through-the-wall": (1) motion detectors, developed by GTRI[37] and Hughes Advanced Electromagnetic Technologies Center

---

[36]J. Altmann, "Cooperative Monitoring of Limits on Tanks and Heavy Trucks Using Acoustic and Seismic Signals—Experiments and Analysis," *Proceedings of the 5th Battlefield Acoustics Symposium*, Ft. Meade, Md., September 23–25, 1997, pp. 135–174.

[37]"Radar Flashlight Illuminates Humans Behind Walls," *Signal*, June 1998, pp. 89–90.

(HAETC),[38] and (2)ultra-wideband radars,[39] which have been developed by Time Domain Corporation, Lawrence Livermore Laboratory, Lockheed Sanders,[40] and others. These devices are being promoted for police work, urban warfare, and medical monitoring. Applications for police and urban warriors include avoiding ambushes and sizing up hostage situations. Through-the-wall radar has promise under some conditions but may prove to be easily countered by simple measures, such as foil lining on walls or even water on walls.

GTRI's motion detector, called a "radar flashlight," is a low-power X-band (near 10 GHz) continuous-wave Doppler radar that fits in a handheld cylindrical package. As with other Doppler radars, it is able to extract a small signal from a moving object in the midst of much higher-amplitude reflections from surrounding clutter. X-band is not optimal for penetrating walls, but the flashlight is sensitive to the movements of human respiration. The shock wave of the beating heart as it propagates to the chest wall—the signal picked up by stethoscopes—can be detected through a 20-cm-thick concrete-block wall.

The current version of the flashlight is designed to be operated while stationary, a limitation that current development efforts are laboring to overcome. As with airborne Doppler radar, the messy problem is how to reject clutter that has relative motion of comparable magnitude to the target. If successful, packaging for a mini-UAV might be feasible.

A motion detector developed by HAETC is a briefcase-sized device operating around 900 MHz. At this low frequency it is able to penetrate through 3 ft of concrete-block wall. The radar does not use Doppler processing but, instead, detects small phase changes when the position of any objects in the room changes. When the received

---

[38]F. Su, "Surveillance Through Walls and Other Opaque Materials," *OE Reports*, No. 140, October 1995, pp. 1–3.

[39]M. Hussain, "Ultra-Wideband Impulse Radar—An Overview of the Principles," *IEEE AES Systems Magazine*, September 1998, pp. 9–14.

[40]"Mini Electronics Smarten Up Small Units," *Jane's International Defense Review*, No. 8, 1998, pp. 34–35; M. Hewish et al., "Ultra-Wideband Technology Opens Up New Horizons," *Jane's International Defense Review*, No. 2, 1999, pp. 20–22.

signal is demodulated, the voltage changes fall into the acoustic frequency band. They are presented to the user as a tone, analogous to submarine sonar, which allows for some degree of target discrimination by an experienced operator.

Ultra-wideband (UWB) radars have been the subject of intense research for nearly a decade.[41] UWB waveforms are either impulse, like the through-the-wall communications technology discussed earlier, or coherently modulated, like spread-spectrum communications. The UWB radars of the impulse variety employ very short pulses of approximately 1-nanosecond (nsec) duration. The coherent modulated waveforms have much longer duration; however, upon demodulation in the receiver, their phase or frequency coding allows them to be compressed to the short duration of the impulse waveforms. Both types of UWB pulses have very high *percentage* bandwidth, which means that the ratio of the frequency spread of the energy to the center frequency is greater than 25 percent. Conventional radars having a low-percentage bandwidth cannot support wideband waveforms at the lower radar frequencies.

UWB radars offer a means to enhance wall penetration *and* range resolution simultaneously. They can operate at the lower radar frequencies that penetrate earth, concrete, etc., with smaller losses than at higher frequencies, and they have the wide bandwidth that is required to obtain good resolution in range. The very fine resolution of UWB radars (typically ≈ 15 cm) also affords relative immunity to multipath, a feature that is held in common with UWB communications.

Both Time Domain Corporation and Sanders have operated UWB radars in the synthetic-aperture-radar mode, obtaining through-the-wall images of objects in a room. The Sanders device, developed under DARPA's Smart Module Program, is called Hand Held Synthetic Aperture Radar (HHSAR). Its 2-GHz bandwidth translates into a resolution in range of 9 cm. Usually, a large enough aperture is synthesized to achieve the same resolution in cross-range as in range. For a room measuring 5 m in depth from the radar aperture, the radar has to be moved laterally a distance of 4 m to accomplish this resolution.

---

[41]OSD/DARPA UWB Radar Review Panel, *Assessment of Ultra-Wideband (UWB) Technology,* Washington, D.C.: Report R-6280, July 13, 1990.

The relative location of the radar across this synthesized aperture must be known to within a few centimeters. Sanders is looking into coupling a miniaturized GPS/Inertial Navigation System (INS) module to the radar for this purpose. Final packaging will determine whether this kind of device is suitable for mounting on robotic vehicles or, perhaps, mini-UAVs. A small package implies that the low-frequency antenna will be inefficient; that inefficiency will be compensated for somewhat by reduced propagation loss through the wall and improved coupling into the target.

UWB SARs with very fine resolution are a technology worth pursuing for urban operations. They can detect and localize (though not identify) individuals inside buildings, which is more instructive than merely detecting the presence of lifeforms. In addition, impulse SARs have the potential for object recognition based on target impulse response.[42]

The idea underlying target impulse response is familiar from acoustics. When a hammer strikes a steel bar, it delivers an impulse, causing the bar to "ring" at its resonance frequencies, which are characteristic of its shape, structure, and composition. The ringing persists after the sound of the initial hammer strike is heard, then damps out. These "late-time" resonances are typically associated with specific scattering centers or modes, such as the propagation of the acoustic wave down the gun barrel and back again some integral number of times.

Irradiating an object, such as a rifle, with an electromagnetic pulse causes the radar echo to display similar tell-tale resonances, if the spectrum of the incident waveform contains significant energy at the resonant frequencies of the object. Typically, an object's lowest frequency resonance occurs at a wavelength twice the length of the object. This amounts to about 2 m for a gun barrel, corresponding to a radar frequency of 300 MHz. For discriminating rifles using reso-

---

[42]C. Baum et al., "The Singularity Expansion Method and Its Application to Target Identification," *Proceedings of the IEEE*, Vol. 79, No. 10, October 1991, pp. 1481–1491; P. Moser et al., "Complex Eigenfrequencies of Axisymmetric Perfectly Conducting Bodies: Radar Spectroscopy," *Proceedings of the IEEE*, Vol. 71, No. 1, January 1983, pp. 171–172; J. Mooney et al., "Robust Target Identification in White Gaussian Noise for Ultra Wide-Band Radar Systems," *IEEE Transactions on Antennas and Propagation*, Vol. 46, No. 12, December 1998, pp. 1817–1823.

nances, we infer that we need to operate the UWB SAR in the VHF band, much lower than the Sanders device. The Army Research Laboratory (ARL) has attempted to detect buried mines using an impulse UWB SAR with its spectral peak in the VHF band—with discouraging results for a small mine buried in loose soil. Results looked promising for larger objects on the surface and closer to a meter in length.

Since SARs can produce an image of the objects in a room, the most effective use of target resonances would be as discriminants for testing objects that appear to be weapons. The imagery would allow the orientation (i.e., compass direction and angle up or down) of the possible weapon to be estimated. Orientation is a required input for estimating the resonant signature. A VHF impulse SAR, if packaged similarly to Sanders' HHSAR, might be an effective ambush detector.

Remote Listening

Remote listening occupies a unique niche: It enables human intelligence to be collected covertly without agents on the ground. This capability could play a role in ambush detection and, in concert with other implanted sensors, in general surveillance.

Laser remote-listening devices were developed decades ago. The concept involves illuminating windows with a low-power continuous-wave laser and recovering conversations in the building from the reflected signal. Acoustic waves impinging on the window cause it to vibrate, much like a microphone diaphragm. This vibration modulates the phase of the laser beam. Of course, unlike the microphone, the window is far from an ideal, high-fidelity transducer.

With the advent of solid-state lasers and miniaturized processors, it is possible to package remote-listening lasers as payloads for mini-UAVs or as implanted sensors.[43] Since the voice bandwidth is narrow, a data link for sending demodulated signals up to a high-altitude UAV need be no larger than a cell phone.

---

[43]J. Anthes et al., "Non-Scanned LADAR Imaging and Applications," *Applied Laser Radar Technology, Proceedings of the SPIE,* Vol. 1936, 1993.

The latest entry into the remote-listening field is a research effort by Sandia Laboratories to develop a *microwave* remote-listening device.[44] Experiment will determine whether microwaves are better able to capture voice modulations from the window than lasers are. One might hope to extend the range of a microwave device beyond the laser; however, if the microwave beam spreads out to cover several windows, overlapping conversations might degrade intelligibility.

Employing higher radio frequencies and millimeter waves instead of microwaves could help. With a typical window spacing of 3 m and a 1-m-diameter antenna operating at 95 GHz, the range could be increased to approximately 1.5 km before two windows are in the beam. To extend the listening range even further, operation could be bistatic, with the illuminator on a UAV and the receiver an implanted or parked device on a facing building. If the implant has a 6-in. antenna and is situated 50 m away, its receive beam would be only 1 m across, the width of a single window. Under remote control, it could interrogate any of the windows illuminated by the UAV's beam and, being passive, would consume very little power, primarily for its data link.

Chemical Sniffing

Several rapidly developing technologies may lead to chemical-/biological-sniffing payloads for mini-UAVs in the near term. Applications to urban operations include detection of explosives (car bombs and weapons caches), mines, drugs, and releases of chemical or biological weapons. Because of this potential for military applications,[45] DARPA is investing heavily in chemical/biological sensing. However, the underlying micro-instrumentation technology is primarily commercial, with applications to medical diagnostics, food processing, chemical industry, hazardous materials ("hazmat")- site profiling, environmental monitoring, cell sorting, protein separation, and DNA sequencing.

---

[44]R. Martinez, Sandia National Laboratories, Albuquerque, N.M., Private Communication, February 13, 1998.

[45]A. Venter, "Trials Planned for Artificial 'Dog's Nose'," *Jane's International Defense Review*, No. 3, 1999.

The most promising of these sensing technologies is microfluidic lab-on-a-chip devices.[46] In general, these consist of an array of microcells or channels—arrays up to 90,000 cells have been developed for genetic research—microfabricated on planar substrates. For detection, each array element reacts with or binds to specific substances. This parallelism enables many reactions to be tested simultaneously, hence the capability for ultra-high throughput screening (UHTS).[47]

Typically, the microchips involve input and output by pipette, inkjet, or electrospray. Chemical separation may be built onto the chip using liquid chromatography or electrophoresis. A variety of techniques have been employed in the final stage of chemical detection, including quadrupole or ion-trap mass spectometry, matrix-assisted laser-desorption ionization time-of-flight spectrometry, cell-based assaying (using multicolor fluorescence analysis in living cells), fiber-optic fluorimetry (the fibers are coated with antibodies aimed at specific biological agents), and electrical resistance of absorptive polymers (the resistance of each polymer changes in a characteristic manner after absorbing the chemical vapor under test).[48]

In 1996, the Naval Research Laboratory (NRL) successfully flew a biological-warfare-agent detection system on a Swallow mini-UAV.[49] The biosensor was of the fiber-optic fluorimetric type. Although not in the form of a microchip, it fit within the 10-lb payload capacity of the aircraft. Further developments in this field are likely to provide even smaller payloads, increased sensitivity, and improved discrimination.

---

[46]R. Marsili, "Lab-on-a-Chip Poised to Revolutionize Sample Prep," *R & D Magazine,* February 1999, pp. 34–38.

[47]T. Studt, "Development of Microfluidic UHTS Systems Speeding Up," *R & D Magazine,* February 1999, p. 43.

[48]J. Hicks, "Genetics and Drug Discovery Dominate Microarray Research," *R & D Magazine,* February 1999, pp. 28–33; S. Henkel, "Tunable Electronic Nose Measures Increased Resistance of Expanding Elements," *Sensors,* March 1999, p. 6.

[49]C. Bovais, "Integration and Flight Demonstration of a Biological Warfare Agent Detection System on an Unmanned Aerial Vehicle," *AUVSI '98,* Association for Unmanned Vehicle Systems International, 1998.

SENSOR FUSION IN SUPPORT OF URBAN OPERATIONS

Sensor fusion is the process whereby information from different sensor types is integrated and presented on a single display. For example, visual, IR, and SAR images can be combined to give a richer picture than any single sensor could provide. Sensor fusion can also be used to reduce false alarms. For example, the sensor-fusion system might be programmed to ignore seismic detections, unless nearby acoustic sensors also detected the unique signature of a vehicle engine.

Effective sensor fusion has proven difficult when more than a few sensors are involved, phenomenologies differ, or the types of errors are significantly different. As well, daunting problems remain in processing, network design for ease of scaling, and algorithm development. As the number of nodes having overlapping coverage in the sensor network increases, the number of operations involved in fusing the information grows extremely rapidly. Achieving a common picture requires the ensemble of sensors viewing each event to be reconciled. This nonlinearity implies that achieving fusion through brute-force processing power is a dubious proposition. The problem is compounded by the tendency to increase the dimensionality of the data to more fully characterize the targets being observed.

Fortunately, short of full-scale data fusion, two steps can provide utility:

1. Assist the operator by collecting multiple sources of information at a single point and allowing the overlay of various data elements. In assisting with visualization, the various sources are largely accepted as ground truth and are usually correlated only within each data type. For instance, if acoustic sensors report three discrete targets and radar reports four targets in the same very small area, the actual count would be ambiguous. It would be difficult to know which, if either, source is reporting the correct number. Consequently, a visualization system might present one set of results, present both sets, or contrive a rough correlation without really addressing some of the ambiguities and resolving the issue fully. Perhaps the greatest value would be in highlighting discrepancies among sensor types.

2. Assist the operator by filtering the data from a large array of sensors so that unusual activities that might warrant a closer look by other sensor systems can be detected and brought to the operator's attention. The system would aid in understanding what is happening by providing, for example, a screen, a monitor that flags unusual events, or a backstop to human observation of the data network. The technology for anomaly detection builds upon the GIS outlined earlier in this chapter. It includes a statistical analysis of both patterns of dynamic activity and unusual changes in objects that can be located again. A now relatively mature technology, anomaly detection can alert human operators of command and control (C2) systems to investigate activities that differ from the norm.

At work in both commercial and military domains, many of the basic technologies for simple fusion are based on neural nets that are "trained" to recognize normal and abnormal activities and to cue humans for intervention when critical thresholds are crossed. Extensive training is required for the neural networks to establish proper baselines of behavior and to establish acceptable false-positive and false-negative rates.

The critical weakness of all current approaches for anomaly detection is that they have a significant high false-positive rate under real-world conditions. Whereas systems that simply look for change can be set to flag changes, dynamic environments require the software to cull abnormal changes from a large array of normal changes that occur on a day-to-day basis. Consequently, these computer-based systems often perform more poorly than a skilled human examining the same data—but they can examine much more data than a human can, and will do so without becoming bored or tired.

Traffic monitoring in a peacekeeping situation is an illustrative application of such a system in an urban area. Here, a large number of sensors might be used to create an estimate of normal traffic patterns as a function of date, day of week, and time of day. By flagging areas of unusually low or high activity, such a system could be used to warn of trouble and could alert controllers to dispatch close-look sensor platforms or patrols appropriately. It could also be used by logistics support groups, operators, and others planning operations in the urban area to better take into account the traffic in and around areas of interest.

AIR-LAUNCHED SENSOR PLATFORMS

As discussed in Chapter Five, air-launched sensors have the potential to greatly improve the ability of manned platforms to detect and identify adversary forces. In some cases, a mini-UAV or air-implanted ground sensor would accomplish this detection by putting an EO/IR sensor close enough to the target to collect high-resolution imagery. In other cases, small ground sensors could be implanted from the air in locations that friendly ground forces did not have access to.

These sensor platforms could be mini- or micro-UAVs, parafoils, other airborne platforms, or remote ground units. In all cases, they would have to be simple—and cheap enough to be disposable, which might limit the type and number of sensors they could carry. Depending on cost and weight, these platforms could exploit the full spectrum of sensor phenomenologies, including acoustic, seismic, EO/IR, magnetic, chemical, and radar.

A variety of mini-UAVs, both battery-powered and gas-powered, with wingspans as small as 4 ft and endurance up to 2 hr, are already flying with small sensor packages. Thus, the technical challenge is less in the design of the aircraft and more in its packaging and deployment. And, as was pointed out in Chapter Five, "Provide Rapid, High-Resolution Imagery for Target ID," a means is needed to quickly get the offboard sensor from the medium altitudes at which manned platforms typically operate at down to its operating altitude of 1,000 ft or lower. A small, light UAV would simply take too long to fly down to its operating altitude.

To solve this problem, an aerodynamic container could be used for a UAV. It would be carried on a hard point on the aircraft exterior or could be deployed through the back right personnel door on the AC-130. It could be guided or ballistic. At the appropriate altitude, it would need to slow and be stabilized in order to deploy the UAV. This might be done with a drogue parachute. The UAV would need folding wings that deployed once it was released from the container. Some aerodynamic and control issues would have to be solved. Once deployed, the UAV would fly autonomously or be remotely piloted to the surveillance area and would broadcast imagery back to the launch platform.

Alternatively, as discussed in Chapter Five, a lifting body/parafoil could carry the sensors, all in a container on the aircraft exterior or launched from the inside of larger aircraft. Some aerodynamic challenges are associated with designing a small lifting body that carries its own parafoil, but, again, the engineering details do not appear excessively demanding. Stabilization and control of the sensor optics could be challenging, since the parafoil may have oscillation problems not encountered on winged air vehicles. These difficulties might be overcome by avoiding dramatic changes in direction. Once the parafoil was established in a constant descending circle over the target area, it should be possible to get a reasonably stable field of view. Clearly, tests with prototype vehicles will be necessary to fully explore these issues. Given the simplicity and light weight of the vehicle, they should not be particularly difficult or expensive.

When more enduring surveillance of a particular building or other site is needed, it is difficult to meet both high-resolution and covertness requirements from airborne platforms. Also, some sensor phenomenologies have such limited range that airborne application is not feasible. Thus, we may want to implant ground sensors from the air for some missions. However, air-implanted ground sensors appear to be more technically challenging than the airborne sensors discussed above. Dating back at least to the Igloo White program of the Vietnam era, the two approaches—high-speed spikes that embed themselves in the ground and parachute packages designed to hang in trees—are designed for rural applications. Urban ground sensors will primarily need to be able to land on and adhere to rooftops, windows, or the sides of buildings. Urban foliage may offer some opportunities to hide sensors, but the sensors would have to be delivered with much greater precision and be much more covert than sensors dropped in vast woodlands or other isolated areas.

Several approaches are possible. In Chapter Five, we discussed implanting a shoebox-sized sensor package by VTOL UAV. The UAV might be able to place the sensor in the optimal surveillance location; however, to maintain covertness most of the time, the sensor would have to be dropped out of line of sight to the target. Even then, there is the possibility that the UAV would be detected acoustically. Once in place, the sensor would need some limited mobility to get to its surveillance location. Although technically feasible, this approach has several weaknesses: requiring a fairly large UAV to hover within a few hundred feet of the target, leaving a detectable object on a roof where it might be discovered, and requiring sufficient mobility to get around and over rooftop obstructions.

Another approach would use a higher-flying manned aircraft, UAV, or munitions dispenser to drop a small, guided sensor. This sensor would fly directly to its surveillance spot, ideally a building wall facing the target, and implant itself, which would have the advantage of minimizing the acoustic signature but the disadvantage of likely discovery. Technical challenges include precision guidance to fly the sensor to within inches or feet of its desired locations to avoid flying through windows; wall-adherence techniques to keep it attached to the building; and resolution so that the sensor would be small enough to avoid casual detection but large enough to see across a street or perhaps farther.

Still another approach, remote ground sensors deployed by agents or friendly forces, has great potential to enhance air operations. Law-enforcement and covert organizations have used such devices to supplement manned surveillance locations for years and have perfected a host of camouflage techniques. Hand-deployed sensors can often be placed very close to the target. Combined with very powerful telephoto lenses and high-quality optics, these systems have the potential to provide imagery of such quality that individuals can be identified—a common requirement in covert and law-enforcement operations.

This successful quality suggests that, as a hedge against the possibility that air-implanted sensors will be infeasible, the USAF R&D community, in concert with other services and agencies, might do well to explore the development of quality remote ground sensors. As well, not only must remote sensors not compromise technologies and techniques essential to other intelligence-collection operations—a key consideration in their development—but means must be developed to limit the consequences of discovery and analysis by adversary technical experts, since any remote sensor has the potential of being discovered. Various self-destruct techniques might be used to prevent the adversary from using or fully understanding key parts of the remote sensor. Since such techniques are never completely reliable, it is likely that the most-sensitive remote-sensor phenomenologies will have to be avoided and less-than-state-of-the-art technologies used in many cases.

LIMITED-EFFECTS MUNITIONS

As discussed in Chapter Four, USAF weapons are optimized for precision attack against medium to hard targets and are extremely valuable in more-conventional urban fights. However, in operations in which restrictive ROE require that damage be limited within buildings, perhaps even to single rooms, these weapons have too much explosive power and penetration potential. Anti-personnel weapons, such as the 40mm and 105mm guns on AC-130 gunships, are more appropriate under these more-constrained conditions, but they also have limitations, particularly against interior targets in urban canyons.

A growing requirement beyond these more-traditional weapons is for highly discriminating weapons whose effects can be tailored to meet the unique needs of each situation. As the precision of air-delivered ordnance has improved over the twentieth century, effects have shrunk from citywide to blocks to individual buildings. It is only natural that airmen would continue this evolution, taking the next step and developing weapons that have effects limited not just to buildings but to individual rooms within buildings: kinetic-energy weapons; laser-guided hand grenades; miniature glide bombs, cruise missiles, and killer UAVs; and nonlethal weapons.

Kinetic-Energy Weapons

As discussed in Chapters Four and Five, several approaches can be taken to make air-delivered ordnance more discriminating in urban settings. The first approach would simply reduce the explosive yield of existing weapons so that the effects would be more limited. Some experimentation would be necessary to understand the effects associated with various smaller warheads. In the extreme, the explosives can be taken out completely, as in the laser-guided training round or the 2,000-lb bombs filled with solid concrete used against Iraqi air defense sites during Operation Northern Watch strikes in October 1999.[50] With explosives removed, the amount of damage is a function of the speed, weight, and density of the weapon casing and fill, variables that can be adjusted for. Such weapons, if sufficiently accurate, can inflict substantial damage against equipment, vehicles, and smaller structures, but their effect on people in structures is harder to predict. In smaller rooms, kinetic-energy weapons are likely to kill or injure occupants. In larger rooms, however, the lethal/injury radius from shock or fragmentation may not cover the entire space. In general, a unitary kinetic-energy weapon is less effective against area targets than is a weapon relying on explosive effects. Additional testing of shock-wave, spalling, and other effects will be necessary to adequately assess the anti-personnel potential of kinetic-energy weapons. Finally, such weapons (at all but the slowest speed/weight combinations) still present a serious penetration danger when damage is to be limited to a single floor, although perhaps they could be designed to shatter on impact to avoid this problem.

"Laser-Guided Hand Grenades"

Alternatively, it may be worth exploring very small, laser-guided weapons, such as the Marines have done with 2.75-in. rockets. A precision weapon of this class, a "laser-guided hand grenade"[51] if you will, could be delivered against targets in all but the steepest and narrowest urban canyons. It would require more-focused and more-precise laser designators than are currently deployed, so that this small weapon would, for example, go through a window rather than bounce harmlessly off the outside of a building.

Miniature Glide Bombs, Cruise Missiles, and Killer UAVs

An alternative to this "hand grenade" approach would use a small, slow-flying platform (such as a UAV or a small cruise missile like LOCAAS) to deliver a small munition (weighing from a few ounces to

---

[50]Selcan Hacaoglu, "U.S. Air Force Using Concrete Bombs Against Iraq," *Associated Press Newswires*, October 7, 1999.

[51]A term coined by RAND colleague David Shlapak a few years ago on a related project.

a few pounds). The main challenges here are developing a platform that is so agile and accurate that it can maneuver down into the urban canyon and either fly by the target and fire a projectile sideways at the target or fly into the target. Lacking much penetration potential by design, both the laser-guided hand grenade and this concept would work best against targets in the open, in open rooms, or behind glass. They would also have to be exceptionally accurate, which is unlikely to be feasible without a navigation system integrating GPS pseudolites and 3-D maps, as discussed earlier in this chapter. This is probably the most technically challenging of the weapon options in this report.

Nonlethal Weapons

Finally, there is the option of using nonlethal weapons against urban targets.[52] These weapons include a wide range of technologies designed to accomplish quite disparate tasks. Their primary attraction for urban operations is the hope of solving the target-discrimination problem by achieving an acceptable effect on adversaries without harming the civilians when combatants and noncombatants are intermingled or are very near by. For example, if a sniper were firing from an apartment building, a nonlethal weapon such as a sedative gas might be used to stop him from firing. If the gas canister missed and landed in someone's living room or if the gas drifted into other spaces, the worst that would happen, in theory, is that the civilians would fall asleep for some period of time. As discussed later in this subsection, there are a variety of reasons why this is very hard to do in practice, but that is the promise.

Most nonlethal weapons are designed for close-in use by infantry or police, but several technologies have promise as air-delivered weapons. Many nonlethal weapons are already being deployed or are in development. The following paragraphs discuss a few of these—acoustic devices, optical effects, nonlethal barriers, high-powered

---

[52]An excellent primer on nonlethal weapons can be found in Robert J. Bunker, ed., *Nonlethal Weapons: Terms and References*, Colorado Springs, Colo.: U.S. Air Force Academy, INSS Occasional Paper 15, 1997; see also John B. Alexander, *Future War: Non-Lethal Weapons in Twenty-First Century Warfare*, New York: St. Martin's Press, 1999.

microwaves, chemical agents, and biological agents—as well as proscriptions against anti-personnel nonlethal weapons.

Acoustic Devices. Acoustic devices, including beams, blast waves, curdlers, squawk boxes, and sonic bullets, are all possibilities for airborne weapons. Some can produce point effects; others produce effects over larger areas. As these technologies evolve, it may be possible to achieve more-precise effects.

Acoustic beams use high-power, very-low-frequency beams to cause body cavities to resonate at particular frequencies. The effects can range from mild nausea all the way to permanent injury and death, depending on range, decibel level, and exposure time. Acoustic blast waves can be generated by pulsed lasers, producing a hot, high-pressure plasma similar to chemical explosives. Acoustic curdling produces a shrieking noise that can be used to disperse rioters. Another crowd-control device that might have utility as a countersniper weapon is the "squawk box," first used by the British Army in 1973 in Northern Ireland. It combines two ultrasonic frequencies that, when mixed in the human ear, produce "giddiness, nausea or fainting." The beam is reportedly so small that it can be directed at specific individuals. Finally, high frequencies can produce an impact wave that hits the target similarly to a blunt object, producing effects ranging from discomfort to death.[53]

The effects of urban structures on acoustic weapons are not completely understood. Some have expressed concern that structures could magnify the effects to potential lethal levels or that, under some combination of high power levels, building-construction materials, and weapon orientations, acoustic weapons could cause structural damage to buildings.[54]

The directionality of weapons effects and range can limit the development of airborne acoustic weapons. To the extent that directionality can be controlled, concepts such as the acoustic beam may be

---

[53]Richard Kokoski, "Non-lethal Weapons: A Case Study of New Technology Developments," in *SIPRI Yearbook 1994*, Oxford, England: Oxford University Press, 1994, pp. 376–377.

[54]Greg Schneider, *Nonlethal Weapons: Considerations for Decision Makers, ACDIS Occasional Paper,* Urbana-Champaign: University of Illinois, 1997, p. 17.

feasible from airborne platforms. If, however, the beam is omnidirectional, it should clearly not be put on a manned platform because it would harm the crew. A UAV might carry such a device, assuming that the acoustic energy would not interfere with or damage the UAV itself. Alternatively, an acoustic-beam-generating device might be dropped by parachute from a manned platform. Some acoustic weapons have ranges measuring a few hundred meters, well below altitudes where manned aircraft typically operate. Even longer range acoustic systems would require manned platforms to fly within the envelope of MANPADS and AAA. For these reasons, it might make sense to put acoustic weapons on low-flying UAVs.

Optical Effects. Optical effects can also be exploited to produce nonlethal effects. Bright lights, strobes, and flash/bang grenades can be used to stun, disorient, or even cause epileptic seizures. For example, high-intensity strobe lights flashing near human brain-wave frequency reportedly cause vertigo, nausea, and disorientation, and might cause epileptic seizures in some people.[55] A less exotic application is found in the Mk-1 illuminating grenade, which was used during the Vietnam War as a counter-ambush weapon. It produced 55,000-candlepower illumination for 25 sec, temporarily blinding those caught nearby. Such devices might also be useful to counter urban ambushes or to prevent a MANPADS operator or sniper from sighting in on his target, allowing the friendly aircraft or personnel to move beyond line of sight. Lasers, such as the Army's Stingray system,[56] can be used to damage optics on sensors and weapons, as well as to temporarily or permanently blind adversary personnel. Airborne lasers might be useful as obscurants, to damage sensors and other optics or to prevent adversary forces from seeing through windows. For example, lasers "have the capability of heating and distorting or cracking the glass lenses of optical systems. This effect is called crazing and is caused when the heat buildup and subsequent cooling in the glass surface creates uneven stresses in the glass surface to crack it. The result is a frosted effect, making it impossible to see through the glass lenses or vision blocks (glass windows) in

---

[55]Artur Knoth, "Disabling Technologies: A Critical Assessment," *International Defense Review,* July 1994, p. 39.

[56]For more on this system, see U.S Department of the Army, FM-90-10-1, 1995, p. J-8.

tanks."[57] Alternatively, an argon laser can be used to temporarily prevent vision through the window of a vehicle or structure. Small abrasions in the glass scatter this frequency of light, causing the entire window to turn an opaque green as long as it is illuminated.[58]

Nonlethal Barriers. Low-friction polymers (super-lubricants), high-friction polymers (sticky foams), aqueous foams, Caltrops (multisided steel barbs), and other devices can be used as nonlethal barriers. Low-friction polymers impede personnel or vehicle movement, producing an ice-slick surface impossible to stand or drive on. Sticky foams produce a gluelike barrier that cannot be penetrated; they were used during the withdrawal from Somalia.[59] Aqueous foams are dense suds that are used in conjunction with barbed wire, Caltrops, and other antimobility devices. The foams make it difficult for adversaries to see and remove the antimobility devices. Caltrops, tetrahedrons, and similar devices are designed to puncture vehicle tires or limit foot traffic. The standard design has four points. No matter how it lands, the device always presents one barb upward. Tetrahedrons (a four-sided barb) were used to interdict North Korean road traffic during that conflict; Caltrops were used by U.S. Marines during the final hours of the withdrawal from Somalia.[60]

Any of these devices might be delivered by air to produce a barrier in a small area, but polymers and sticky foams are best delivered in urban areas by ground vehicles or stationary equipment. In situations where a small area—such as a rooftop, alleyway, stairway—needed to be blocked, one could imagine a UAV, LGB, or glide bomb delivering foam or low-friction polymer. Caltrops, in contrast, could be easily delivered over a large area by aircraft.

High-Powered Microwaves. High-powered microwaves (HPM), which are transmitted by a radarlike antenna or generated through an explosive device, have potential as air-delivered nonlethal weapons for both anti-personnel and antimateriel applications.

---

[57]Bunker, 1997, p. 16.

[58]Bunker, 1997, pp. 17–19.

[59]Bunker, 1997, p. 8.

[60]See Robert F. Futrell, *The United States Air Force in Korea: 1950–1953*, Washington, D.C.: Office of Air Force History, 1983, p. 328; and Frederick M. Lorenz, "'Less-Lethal' Force in Operation United Shield," *Marines Corps Gazette*, September 1995, p. 74.

Some have advocated HPM weapons in the anti-personnel role because of their supposed potential to render personnel unconscious without permanent damage.[61] Yet, given what is known about the effects of microwaves on humans, it seems unlikely that these weapons could be that benign.

HPM-induced changes in brain temperatures of a few degrees (in laboratory rats) caused convulsions, unconsciousness, and temporary blindness.[62] Higher dosages on humans could cause effects ranging from heart and respiratory failure to permanent brain damage. The power density required to produce unconsciousness in humans is "10 to 50 milliwatts per square centimeter for continuous exposures at microwave frequencies up to 10 GHz . . . [while the] . . . single-pulse fluence that produces significant heating at these frequencies is about 100 joules per square centimeter."[63] This is much higher than what is required to damage electronics. For example, some microwave detector diodes will burn out at 1 microjoule.[64] Thus, in theory, antimateriel weapons could be made to preclude the worst effects on humans. Whether this could be done in practice remains unclear, since the pulse degrades with range. To produce a pulse that would damage electronics in a target 1 km away might require power levels that would harm humans closer to the HPM source.

HPM weapons create an electromagnetic pulse that produces a surge of power through unprotected electrical equipment, potentially disabling vehicles, radios, computers, and radars. Depending on the power levels experienced by the target, the damage may be transitory (e.g., requiring computers to be rebooted) or permanent (e.g., by physically damaging integrated circuits). Designed without significant protection against low-power accidental interference, commercial systems (especially communications) are usually more vulnerable to these type of effects. Consequently, if the military mission required that an adversary's systems be permanently damaged,

---

[61]Kokoski, 1994, p. 374.

[62]H. Keith Florig, "The Future Battlefield: A Blast of Gigawatts?" *IEEE Spectrum,* March 1988, pp. 53–54.

[63]Florig, 1988, p. 53.

[64]Florig, 1988, p. 53.

the higher power required to do so would increase the chances that nearby civilian systems would be damaged also. This might limit the use of HPM near essential civilian electronics (e.g., telecommunications, hospitals, electrical-power facilities) and might rule out its use where civilian and adversary systems were located in the same building. In most cases, however, it appears that HPM can be tailored so that permanent damage is limited to quite small areas.

Chemical Agents. A variety of reactant chemical agents have been developed as antimateriel weapons. These include combustion-altering agents, super-caustic agents, and liquid-metal embrittlement. Some of these agents would be quite lethal if humans were exposed; their nonlethality assumes that humans are not nearby when they are used.

Combustion-altering agents either contaminate or change the viscosity of fuel, causing engine failure. They can be delivered as a vapor through engine air intakes or introduced into the fuel supply. Super-caustic agents are mixes of acids that will dissolve most metals. They could be used against buildings, roads, and vehicles. Liquid-metal embrittlement changes the molecular structure of base metals, potentially causing structural failure of bridges, buildings, aircraft, and ground vehicles.[65]

Most of these agents could be delivered by air. However, their greatest utility is for special operations rather than for routine use by general-purpose forces. The political consequences of causing civilian injuries with super acids or other volatile compounds could be devastating in more-constrained operations and will likely keep these compounds from becoming widely used in urban settings.

Biological Agents. Finally, a variety of biological and chemical nonlethal agents are available, such as tear gas, calmative agents, malodorous agents, and sickening agents. The practicality of these concepts varies, but as is discussed in the next subsection, we do not believe these weapons have much applicability in urban operations.

Proscriptions Against Anti-Personnel Nonlethal Weapons. Nonlethal weapons clearly have utility in some urban military situa-

---

[65]Kokoski, 1994, p. 377.

tions, particularly those faced by special operations forces. However, several factors are likely to prevent anti-personnel nonlethal weapons from being widely used when civilians and adversary forces are intermingled.

First, nonlethals fail to meet mission requirements in many situations. Most of the time, U.S. forces want to permanently remove adversary forces from the fight by capturing or killing them or to undermine an adversary's morale by producing casualties. Consider an adversary sniper firing on friendly forces. Knocking him out with a nonlethal weapon would have the benefit of stopping him from harming any other friendlies at that time. Yet, unless friendly forces were able to find and capture the unconscious sniper, he would be fit to return to fighting soon thereafter. This also would have the undesirable effect of under-deterring violent actions. Also, not all non-lethal-weapon effects occur immediately. Timing the onset of effects and limiting their duration can be quite complex; one could not be certain that the adversary was incapacitated at the critical time, which suggests that, most of the time, lethal weapons would be the preferred option.

Second, there is the possibility that nonlethals would, in fact, kill or permanently harm civilians. Urban areas increase the probability that nonlethals could harm civilians because (1) the high population densities simply increase the number of people who might be exposed to an amount, or in a way, that would be harmful and (2) enclosed spaces, whether alleys, courtyards, or interior spaces in buildings, may interact with nonlethal weapons in unforeseen ways to concentrate dosages, intensify effects, or limit avenues of escape.

Nonlethals are attractive because they might solve the target-discrimination problem when adversary forces and noncombatants are intermingled. A riot-control agent or acoustic weapon might be used to drive an intermingled group away or to incapacitate them until friendly forces could sort them out on the scene. Technologies already exist to drive people off with fairly low risk. But rendering people unconscious is a much trickier business; it has the potential of killing the young, old, or sick, or of causing permanent harm. Each of these concepts will need to be explored in great depth to ensure that the effects are sufficiently benign to use against noncombatants. However, if the alternative is firing lethal weapons into a crowd, such risks might appear small. The difficult question for policymakers is whether there are a sufficient number of such circumstances to justify developing and deploying these systems as backups.

Third, the U.S. is signatory to a number of agreements that may prohibit or limit the use of some nonlethal weapons. For example, states that signed the Biological Weapons Convention of 1972 agreed not to "develop, produce, stockpile or otherwise acquire or retain . . . microbial or other biological agents, or toxins whatever their origin or method of production, of types and in quantities that have no justification for prophylactic, protective or other peaceful purposes."[66] Many of the biotechnical concepts appear to run afoul of this agreement. The Chemical Weapons Convention of 1993 obligates (in Article I) signatory nations not to use chemical weapons, which it defines (in Article II) as "any chemical which through its chemical action on life processes can cause death, temporary incapacitation or permanent harm to humans or animals." It specifically states that "each State Party undertakes not to use riot control agents as a method of warfare." [67] Finally, both customary international law and treaty law restrict the use of weapons that cause superfluous injury or are entirely incapable of discrimination. To the extent that weapons such as lasers, high-power microwaves, or acoustic weapons produce long-term health problems, they might be arguably in violation of this principle. In sum, a strict reading of these agreements and customs could rule out the use of many of the nonlethal concepts being explored today.[68]

Even if the use of certain nonlethal weapons is not prohibited or limited under international law, the public reaction—in the local urban setting, the region, and globally—to their use could produce costs that greatly exceed any immediate tactical advantage. Even relatively benign weapons such as CS gas[69] could produce lethal effects in enclosed spaces and against the young, elderly, and sick. It could also

---

[66]http://www.acda.gov/treaties/bwc1.htm

[67]http://www.acda.gov/treaties/cwcart.htm#I

[68]For more details on the potential implications of these treaties for nonlethal weapons technologies, see Barbara Hatch Rosenberg, "'Non-lethal' Weapons May Violate Treaties," *The Bulletin of the Atomic Scientists,* September/October 1994, pp. 44–45.

[69]CS gas is the most widely used riot-control agent.

cause a crowd to panic and stampede or crush people to death in a rush to escape the gas. Adversary propaganda, local myths, and rumors could cause overreactions through misinformation about the agents (e.g., claiming that they were lethal, caused infertility, or carried other frightening effects).

Although, in many cases, the effects from nonlethals are less harmful or at least no more harmful than those of conventional weapons, the nature of the effects could produce disastrous political fallout. For example, imagine an operation that resulted in civilians being blinded or terribly burned with acids instead of being killed. The media and public reactions to the blinding and burning events would likely be much worse, at least in part because there would be survivors to photograph, interview, and write about. Therefore, they are likely to be a factor of great concern to U.S. leaders anytime the use of these weapons is contemplated.

In short, we recommend continued research and development of nonlethal weapons for appropriate situations but see them as having little applicability in most urban operations. For this reason, this study has emphasized more-conventional weapons.

CONCLUSION

This chapter has sought to show the breadth of technological developments relevant to urban aerospace operations. Many technical hurdles remain and some capabilities may be decades from being realized. However, many technologies are sufficiently mature to justify the development of prototypes and the initiation of operational testing. As noted at the beginning of this chapter, these systems have the potential to greatly enhance aerospace operations in urban environments, but they are unlikely to come to fruition without strong institutional support to take promising ideas out of the laboratory and into the field.

We now turn to Chapter Seven and offer some final observations about the role of aerospace forces in urban settings.

Chapter Seven

CONCLUSIONS

INTRODUCTION

This report has sought to provide an overview of the varied challenges—from legal constraints to line-of-sight limitations imposed by urban geometry—facing airmen in urban military operations, and of operational concepts and new technologies for dealing with those challenges. Although the urban setting is a complex and difficult environment in which to operate, aerospace forces can make important and unique contributions to joint urban operations. We conclude this report with a short summary of our key findings, some thoughts on the importance of urban operations, and some caveats about the performance of sensors, weapons, and people in actual combat. Finally, we recommend some steps the USAF can take toward acquiring the capabilities discussed in this report.

KEY FINDINGS

Key findings of this study are as follows:

- Global urbanization, particularly in the developing world, makes it highly likely that many, if not most, future military operations will have an urban component (although not necessarily one involving fighting).

- An increase in urban operations does not mean that conflict has become primarily an urban phenomenon or that nonurban military operations have been eclipsed. Rather, built-up areas are yet another environment in which military forces must be prepared to operate.

- The physical and social complexity of urban areas makes them extremely difficult to operate in. Where possible, U.S. forces should avoid them. Aerospace forces can help preclude some urban military operations through deterrence, early warning, and rapid humanitarian or military intervention. Along with ground-based long-range fires, they can interdict adversary forces, potentially preventing them from reaching urban areas.

- Where urban operations cannot be avoided, aerospace forces can make important contributions to the joint team, detecting adversary forces in the open; attacking those forces in a variety of settings; and providing close support, navigation and communications infrastructure, and resupply for friendly ground forces.

- Offboard sensors for manned aircraft, three-dimensional urban mapping, GPS relays on UAVs, and limited-effects munitions have the potential to enhance the ability of aerospace forces to detect and attack adversary forces where rules of engagement are highly restrictive, such as in peace operations, noncombatant evacuations, and humanitarian assistance. Their development should be encouraged.

- Three-dimensional mapping and GPS relays also have the potential to substantially improve the situational awareness of friendly ground forces, allowing the smallest units, as well as their commanders, to know their own location (both GPS coordinates and position in buildings). Coupling these technologies with laser rangefinders should allow friendly forces to quickly map the location of engaged adversary forces.

- Automated integration and pattern analysis of inputs from large, multiphenomenology sensor networks will be necessary to make sense of the massive volume of activity found in most urban areas.

- But, in the type of limited operations this report emphasizes, it is *unlikely* that automated classification of weapons, adversary personnel, or vehicles will be sufficiently reliable to permit lethal fires to be put automatically on targets. Rather, we expect that practical limitations of automated fusion, coupled with political concerns about collateral damage and civilian casualties, will dictate that at least one human decisionmaker remain in the loop between sensor and shooter.

- As long as human decisionmakers remain in the loop between sensor and shooter, human-machine interfaces will be a critical information-architecture issue. A major challenge will be developing the organizational processes that make quick decisions possible in light of the likely uncertainty and ambiguity associated with real combat. Without a responsive and agile command and control system, an elusive and adaptable adversary is likely to be there and gone before weapons can be brought to bear.

THE NEED FOR IMPROVED URBAN OPERATIONAL CAPABILITIES

Are urban-centered conflicts becoming more common? Are they a new form of warfare that will supplant traditional maneuver warfare in the open? These are intriguing questions that deserve serious and careful consideration by defense planners and researchers alike. At this point, there is insufficient evidence or analysis to answer them. In finding answers, defense planners must walk a narrow path between apocalyptic and complacent visions of the future security environment. They should focus, at least for now, on ensuring that the U.S. military can meet a broad range of urban-military-operation challenges, whether in major wars or small-scale operations.

The best argument for improved urban-military capabilities is that, despite its best efforts to avoid them, the U.S. military has had to fight in cities in a multitude of circumstances. And, in an increasingly urbanized world, noncombatant evacuations, humanitarian relief, and other "noncombat" operations are likely to take place in urban settings. The possibility of armed interference in many of these operations means that the military is tasked. Although U.S. forces have usually been able to avoid combat during these operations, they must be prepared to conduct urban evacuations and humanitarian relief in the face of armed opposition. In short, whether in conventional conflicts or in smaller-scale contingencies, there is a good chance that U.S. forces will be called upon to operate in urban settings. Prudent defense planning requires that we develop the doctrine, training, organizations, equipment, and concepts of operations to be effective in this unique and difficult environment.

Planners also need to distinguish clearly between the problem of conducting military operations in the midst of a civilian population and that of fighting in the rubble of a largely abandoned city. The former, more-complex problem deserves analytic attention. As DoD places more emphasis on stability, relief, counterterrorism, and other operations at the lower end of the conflict spectrum, planners, operators, and analysts need to gain a deeper understanding of the human and physical intricacies of the urban environment.

TECHNOLOGICAL PROMISES AND THE REALITY OF WAR

The concepts presented in this report have great promise, but we do not want to imply that the concepts or enabling technologies are a panacea that will make urban operations easy or guarantee U.S. dominance of the urban environment. The urban environment is too complex to lend itself to a simple technological solution.

Even with an extensive aerospace-ground sensor network, much adversary activity is likely to go undetected. Smoke (from fires and adversary smoke-generating machines), dust, inclement weather, night, electronic interference, building materials, and human activities will hinder intelligence collection at one time or another. An adversary force's efforts at deception will, at times, confuse and confound U.S. attempts to detect and attack them. Adversary forces that are detected by ground-sensor fields may move out of sensor range or line of sight before they can be positively identified or attacked. Airborne surveillance platforms at times may be too far off to assist or may lose line of sight at critical moments. Around-the-clock airborne fire support, which is feasible but demanding for platforms and crews, does not ensure that a weapon can be delivered anywhere in the city at a moment's notice. In many cases, adversary forces are likely to escape before the sensor-controller-shooter loop can be completed. A highly active adversary might swamp controllers with alerts, hampering the time-consuming task of visually identifying the targets. Or it might not be possible to identify the targets as hostile because the targets are intermingled with noncombatants.

There also will be occasions when weapons will be released but will miss the intended target because of these factors or adversary countermeasures. At least some of the time, both lethal and nonlethal weapons will injure and kill noncombatants and friendly forces. Limited-effects munitions are also likely to be insufficiently lethal in some situations, allowing adversary forces to escape unscathed or with lesser injuries than they would have received from traditional weapons. Thus, although the weapons proposed here can substantially lessen unintended damage, they will at times fail to achieve the desired effect.

Finally, it must be recognized that the adversary is a thinking, adapting, often highly motivated independent actor who will do creative and surprising things to counter U.S. sensors, weapons, and concepts of operation. Concepts of operation will have to be flexible and evolve to stay one step ahead of such a thinking adversary.

The greatest advantage U.S. forces have over potential adversaries is their ability to integrate air, land, sea, and space forces. In the urban setting, highly integrated operations could allow a ground patrol to send GPS coordinates and/or an image of an adversary's position to a combat aircraft overhead. Alternatively, a UAV, manned aircraft, or satellite might send GPS coordinates, an image, or other data about a target that was around the corner or otherwise beyond line of sight to a ground unit. Mutual sharing of information and images could dramatically increase the effectiveness of both aerospace and ground forces as they gained a perspective on a developing situation impossible to achieve from either the ground, air, or space alone. Integrating inputs from ground, air, and space sensors, along with human observations from patrols or forward observers, could also give commanders a rich, multidimensional view of the battlespace, improving the quality and speed of force allocation and other critical command decisions.

Fully integrated air-ground operations that capitalize on the unique strengths of ground and air forces should minimize U.S. susceptibility to an adversary's deception efforts and risks to U.S. personnel while maximizing the effectiveness of U.S. forces in detecting, identifying, and neutralizing adversary forces.

NEXT STEPS

The caveats discussed above notwithstanding, the general approach presented here is sound and, if pursued, will yield a significant improvement in USAF and joint capabilities for urban operations. In particular, the integration of ground-sensor networks, low-flying air-launched UAVs, and more-traditional surveillance platforms with platforms carrying limited-effects munitions will make it possible for aerospace forces to greatly increase the contribution they make to joint urban operations. Developing the ability to detect, identify, and neutralize room-sized targets without collateral damage is a logical step in the evolution of aerospace power, simply continuing current trends in C3ISR, battle management, and precision strike.

For these capabilities to be realized, several areas will require more-focused attention:

- Air-launched offboard sensors

- Limited-effects munitions and associated platforms

- Non-imaging sensors for ground networks (particularly weapon-detection and explosives-detection technologies)

- Three-dimensional mapping and databases

- Sensor fusion

- Joint command and control of aerospace and ground forces.

In view of budgetary realities and current modernization priorities, we recognize that funds available for enhancing USAF urban capabilities are limited. USAF R&D is already directed at certain capabilities that would be useful in urban settings, such as loitering sensors/platforms and directed-energy weapons. However, programs would have to be initiated or redirected to develop other key urban capabilities. For that reason, we recommend that the USAF continue modest research to identify the most-promising and versatile technologies for urban settings. Additional research and testing will have to be done before there is sufficient data on performance and cost for the USAF leadership to make informed decisions on whether to field systems such as those discussed in this report.

For the near term, we recommend that one of the USAF major commands or a battle laboratory be given responsibility for conducting additional research and development of these systems. To make the most of limited R&D funds, USAF laboratories should seek to partner with the Defense Advanced Research Projects Agency (DARPA) and other interested parties—perhaps under the auspices of an Advanced Concept Technology Demonstration (ACTD) program—to build and test prototypes of the more promising systems. U.S. allies are likely to be important players as well, having developed a variety of nonlethal, countersniper, and other systems that can be applied in urban operations.

Ultimately, urban operations are a joint problem. Theater commanders, the joint staff, and DoD will have to determine which mix of capabilities offers the most robust force for urban operations. Specific sensor and weapon choices will have to be made on the basis of some combination of coverage rate, resolution, versatility, responsiveness, cost, proportional/adjustable effects, and ease of delivery. Urban-specific MOEs may be needed to evaluate options for accomplishing the various tasks. As promising technologies are identified, realistic field testing, simulation, modeling and red-teaming will be necessary to determine which, if any, of these are sufficiently robust under actual operational conditions to justify fielding.

Likewise, certain capabilities discussed in this report, such as urban pattern analysis and the fusion of aerospace-ground sensor inputs, should be developed under the auspices of joint initiatives. These capabilities belong in a joint fusion or command center.

Indeed, some of the most difficult issues are related to joint command and control of urban operations. For example, coordinating joint fires to prevent friendly forces from firing on one another will become a bigger problem if the number of standoff weapons used in urban operations is increased. If significant numbers of friendly forces are on the ground, will all urban air strikes be considered CAS? Or will aerospace forces operate more autonomously in some parts of the city? These are just a few of the many issues that need to be resolved before highly integrated urban aerospace-ground operations become feasible.

We also recognize that other potential aerospace applications may prove to be too expensive or too far removed from the core responsibilities of the USAF to justify the diversion of resources. However, it would be unfortunate if excessive concern about budgetary constraints, combined with somewhat outdated views of the limitations of aerospace power, prevented promising new capabilities from being fielded. Ironically, airmen are often as likely as infantrymen to narrowly define the settings in which aerospace forces can contribute. A more expansive vision of aerospace power would see the urban canyons of the world as part of the continuum of the third dimension that runs from the ground to orbital altitudes. It would embrace nontraditional systems—such as air-dropped UAVs—as simply another tool in the airman's kit bag. The USAF excelled during the twentieth century at going higher, faster, farther. To meet the challenges of the early twenty-first century, the USAF may also need to exploit unmanned and robotic systems so that it can go lower, slower, and closer against unconventional threats to U.S. interests.

TRIGONOMETRIC CALCULATIONS
FOR URBAN LINES OF SIGHT

This appendix explains the trigonometric calculations behind much of the information presented in Chapter Four.

Figure A.1 gives an example of how to determine the maximum horizontal distance, line AD, a UAV can be away from a street and still see three-fourths of that street over a building of a given height.

RAND *MR1187-A.1*

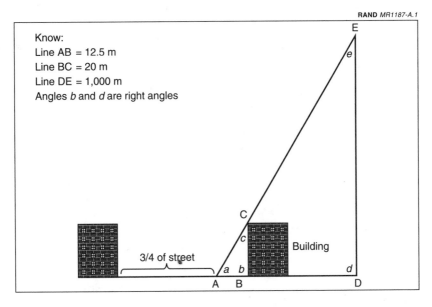

**Figure A.1—Determining Maximum Horizontal Distance
for Viewing Streets over a Building**

The figure uses the average building height of 20 m and street width of 50 m given in Table 4.2 for UTZ V, to illustrate the computation. To be able to see three-fourths of the street means that line AB is one-fourth of 50 m, or 12.5 m, that will not be seen. Line BC is the average building height, 20 m. The UAV altitude, defined as line DE, is 1,000 m, and angles b and d are right angles. With this information, we can use simple trigonometry to find the length of line AD, as follows:

First we find the tangent of angle a. The tangent of angle a is the height of triangle ABC divided by its base. In this case, this is the 20-m building height divided by one-fourth of the street width, 12.5 m.

Since angle a is the same in triangles ABC and ADE, we know that the tangent of angle a must also equal the UAV altitude (DE) divided by the maximum horizontal standoff (AD), or 1.6:

$$\tan a = \frac{BC}{AB} = \frac{20 \text{ m}}{12.5 \text{ m}} = 1.6 \tag{A.1}$$

Simple algebra shows that AD equals 625 m:

$$\tan a = 1.6 = \frac{DE}{AD} = \frac{1,000 \text{ m}}{AD} \tag{A.2}$$

$$1.6 = \frac{1,000 \text{ m}}{AD} \tag{A.3}$$

$$\frac{1,000 \text{ m}}{1.6} = AD \tag{A.4}$$

$$AD = 625 \text{ m} \tag{A.5}$$

The arctangent function on any scientific calculator (or table) can be used to find that angle a is about 58°. This is the minimum angle at which the UAV can see three-fourths of the street. If the UAV moves closer than 625 m, it will be able to see more of the street; as it moves farther away, it will be able to see progressively less of the street.

Similar calculations form the basis of the information presented throughout Chapter Four.

MICROWAVE RECHARGING OF MINI-UAVs AND MICRO-UAVs

A major drawback to employing very small, remotely piloted aircraft in certain scenarios in urban operations is that they may not have sufficient range or endurance to be recovered. Clearly, this may not be an issue if there are sanctuaries controlled by friendlies within the city, or not too far outside. The question we address in this appendix is whether it is feasible to reuse mini–unmanned aerial vehicles (mini-UAVs) or micro–aerial vehicles (MAVs) by recharging them, using either solar energy or microwave beams directed downward by high-altitude UAVs.

The assumed characteristics of the UAVs are listed in Table B.1.[1, 2] The average power calculation assumes a mix of 80 percent level flight and 20 percent maneuver flight.

The approximate ratio of the weights of the two aircraft featured in the table is nearly 100, yet, surprisingly, the average electrical power expended per unit area of wing surface is only 20 percent lower in the micro-UAV. For both platforms, interestingly, the power density is somewhat less than the maximum irradiance of the sun, i.e., 0.137 W/cm^2. If solar panels were able to transform solar photons to electricity with at least 60-percent efficiency, the aircraft could fly on

---

[1]W. Davis, Jr., et al., "Micro Air Vehicles for Optical Surveillance," *The Lincoln Laboratory Journal*, Vol. 9, No. 2, 1996, pp. 197–214.

[2]M. Cantella, "Micro Air Vehicle Sensor," *Proceedings of the IRIS Specialty Group on Passive Sensors*, Vol. 1, 1999.

Table B.1

Sample Characteristics of Mini-UAVs and MAVs

	Mini-UAV	MAV
Weight (g)	4540	49
Wingspan (cm)	121	15
Aspect Ratio	7	3
Wing Area (cm^2)	2096	76
CL	0.6	0.6
CL/CD	15	5
Propeller Efficiency	0.8	0.5
Electrical Efficiency	0.6	0.6
Average Power (W)	178	5.1
Average Power/Wing Area (W/cm^2)	.085	.068

NOTE: CD = aerodynamic drag coefficient; CL = aerodynamic lift coefficient.

solar power with the sun directly overhead. For cases when the sun is not directly overhead, we explore alternatives.

It seems unlikely that the mini-UAV can land in hostile areas for recharging, survive, then take off for another mission. If it can reach sanctuary, simple refueling of an internal combustion engine is the most efficient approach.

The MAV is designed to perch on buildings, with some chance for covertness. Recharging in this instance is practical. Assuming that the solar panels recharge the MAV's battery with an overall efficiency of 20 percent, 2.5 hr of sunlight would be required for the MAV to fly for 1 hr. A critical issue is whether the solar panels can be made light enough for the MAV to carry them.

Another alternative means of recharging the MAV is by beaming down microwaves from a Global Hawk–class or larger UAV at 60,000 ft altitude. The power per unit surface area obtained as a function of radiated power is shown in Figure B.1. MAV P(avg) is the average power required for the MAV to fly. There are separate curves for radar frequencies of X-band (10 GHz), Ku-band (18 GHz), and the millimeter-wave frequencies of 35 GHz and 95 GHz. With the size of the antenna assumed fixed at 24 ft^2, the antenna gain and effective radiated power increase as the square of the frequency. Therefore, the power densities depend on frequency.

Supposing the conversion efficiency for microwaves to electricity is higher than for solar power, around 90 percent, a 100-kW radiator beaming down through all the daylight hours at 95 GHz would be required to recharge a MAV for 1 hr of flight. The radiated power is a factor of several times what Global Hawk can deliver in *prime* power, and the efficiency of antennas at 95 GHz is much worse than at X-band. Moreover, at this high frequency, the beam is very focused, and the MAVs undergoing recharging would be confined to the area of a city block (<2000 m^2).

In summary, whereas recharging mini-UAVs is not practical, refueling them may be, under some circumstances. Solar recharging of MAVs is practical, provided that flight can be restricted to roughly 1 hr out of every 3.5 hr and that solar panels that are very light, yet efficient, become available.

RAND *MR1187-B.1*

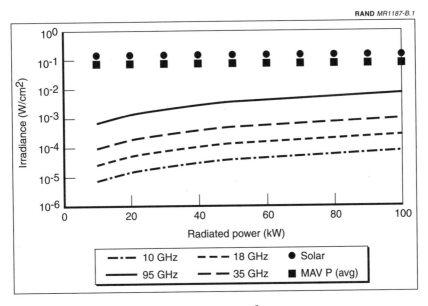

Figure B.1—Irradiance from a 24-ft^2 Antenna at 60,000 ft

DETECTING SNIPERS

A representative selection of the most advanced countersniper systems available in the Free World, and the phenomena they are designed to detect, is shown in Figure C.1. All the manufacturers are located in the United States or Western Europe. The phenomena detected include muzzle blast and flash; the shock wave, vortex, and thermal signature of the bullet in flight; and retro-reflection from the sniper's optical sight.

Muzzle blast and *flash* are the acoustic and infrared (IR) signatures associated with the ejection of the bullet from the sniper's rifle.[1] The muzzle blast can be detected with acoustic sensors at ranges from several hundred meters out to more than a kilometer. The muzzle flash can be detected with IR sensors out to a kilometer or more, but the sensors must have line of sight to the weapon, and the flash can be suppressed.

The *bullet's shock wave* is a mini–sonic boom resulting from the bullet traveling at speeds faster than sound.[2] It can be detected acoustically at ranges from hundreds of meters out to more than a kilometer. If the sniper uses a fire suppressor to slow the bullet to subsonic speed, the acoustic signature of the bullet in flight is hard to detect.

---

[1] S. Moroz et al., "Airborne Deployment of and Recent Improvements to the Viper Counter Sniper System," *Proceedings of the IRIS Specialty Group on Passive Sensors,* Vol. 1, 1999, pp. 99–106.

[2] L. S. Miller, "Counter Sniper Technology," *Proceedings of the 5th Battlefield Acoustics Symposium,* Ft. Meade, Md., September 23–25, 1997, pp. 681–692.

However, this countermeasure reduces the sniper's ability to penetrate armor.

Like most aerodynamic bodies, the bullet sheds vortices in flight, creating disturbances in atmospheric pressure along its trajectory. These vortices produce gradients in the atmosphere's refractive index that can be detected, in principle, with laser radars. None of the systems in Figure C.1 is designed to detect this signature.

The thermal signature of the bullet in flight can be detected with IR sensors out to several kilometers in range. Since the bullet is much hotter than "room temperature," it is detected most effectively in the medium-wave infrared (MWIR) band, with wavelength between 3

RAND *MR1187-C.1*

Name	Manufacturer	Muzzle Blast	Bullet Shock Wave	Muzzle Flash	Bullet in Flight (IR)	Optics Laser Reflection
Prototype	Sanders	X	X			
Bullet Detection Indicator	GD Associates		X			
Bullet Ears	BBN	X	X			
PD Cue	AAI Corporation		X			
Pilar	Metravib	X	X			
VIPER	Maryland Advanced Development Lab			X		
Prototype	Hughes Aircraft	X			X	
Integrated Sniper Location System	Sanders, LMIIS, and Sentech	X	X		X	
Sight Laser Detector (SLD)	Cilas					X
Target Observation and Locating System	Sanders					X
Sniper Acoustic Detection Sensor	Rafael	X				
SECURES	Alliant Techsystems	X				
Sentinel Sniper Location System	SAIC	X	X			
Fast IR Sniper Tracker	Thermo Trex			X		
Lifeguard	LLNL				X	

Figure C.1—Free-World Countersniper Systems

and 5 μm. However, long-wave infrared (LWIR)-based systems operating in the wavelength band between 8 and 10 μm—e.g., the Integrated Sniper Location System prototype—can also detect such signatures.

The object of detecting signatures of the bullet in flight is to estimate the bullet's trajectory and backtrack to find the location of the sniper. Acoustic sensors are passive. Taken singly, they can measure angles to the acoustic source, but not the range. To establish a track of the bullet requires that an array of acoustic sensors be deployed.[3] One alternative approach is to obtain an approximate direction to the sniper from the acoustic information, then to cue an IR sensor to backtrack the bullet more precisely. A second alternative is to detect the muzzle flash with a wide-field-of-view IR sensor, which then initiates an IR track of the bullet, resulting in a backtrack to the sniper.

Urban noise and glare make all of the systems that depend for initial cues on muzzle flashes or blasts subject to high false-alarm rates. For this reason, there is a trend toward multiple-phenomenology systems, which look for coincident detections of acoustic and IR events.

The backtracking process in the city is complicated by buildings, which may obstruct the view of the sniper's window. If much of the bullet track is visible, it is feasible to use the urban models discussed in Chapters Five and Six to complete the backtrack in the virtual world of the computer. This procedure could provide GPS coordinates for a weapon delivered from a UAV.

Laser systems that illuminate potential hiding places, or "hides," and detect retro-reflections from the sniper's scope are referred to as *optical augmentation systems*. These systems have the advantage of possibly detecting the sniper before he fires his weapon. The downside is that the sniper can employ antireflection filters that selectively block the wavelength of the laser. Tunable lasers may reduce the effectiveness of blocking filters in the future.

---

[3]E. Page, "The SECURES Gunshot Detection and Localization System, and Its Demonstration in the City of Dallas," *Proceedings of the 5th Battlefield Acoustics Symposium*, Ft. Meade, Md., September 23–25, 1997, pp. 693–716.

LESSONS LEARNED FROM PAST
URBAN AIR OPERATIONS

This report has investigated the conceptual, legal-political, physical, and technological underpinnings of present-day urban air operations. This appendix looks at the historical record to determine the role that aerospace forces have played in past urban battles, the tasks that they have been assigned, and the conditions that have contributed to their effectiveness or ineffectiveness. The result is an overview of urban air operations from World War II to Bosnia, focusing on battles in which a major purpose of aerospace power was to assist friendly ground forces and/or civilians in contested urban areas. We emphasize U.S. air operations but have sought to learn from any air force that conducted urban operations. The intention is to cover a range of operational examples from urban warfare to military operations other than war (MOOTW); include both successful and unsuccessful urban operations; incorporate a variety of aerospace power tasks; and examine the employment of fixed- and rotary-wing aircraft in an urban environment.

This appendix concludes that all four U.S. military services have accumulated considerable experience in providing air support to joint urban operations during periods of war and relative peace. Despite this extensive record, the effectiveness of U.S. aerospace power in urban operations has varied so much throughout the years that no general trend is discernible. Furthermore, although this appendix analyzes the circumstances where aerospace forces have and have not been effective, the wide array of past examples of urban operations makes it impossible to offer a formula for aerospace force success that would fit the majority of cases. That said, we should note that the United States has not fought in a major urban battle since the real revolution in aerospace power occurred in the late 1980s and early 1990s. The capabilities demonstrated in Operations Desert Storm, Deliberate Force, and Allied Force—most notably the combination of battlefield intelligence collection, stealthy platforms, and precision munitions—would likely make aerospace forces much more effective in large-scale urban operations against conventional foes.

The appendix is divided into three main sections: close air support, air logistics support, and air interdiction and siege support. Each of these sections initially describes the performance of aerospace power in the given functional role and subsequently analyzes the factors contributing to aerospace power's success or failure. In the final section, some observations are made regarding the overall effectiveness of U.S. aerospace power in past urban operations with the hope that these insights will be useful for planning the use of aerospace forces in future urban operations.

CLOSE AIR SUPPORT

From Stalingrad to Grozny, close air support has compiled a mixed record of achievement in urban operations. Historically, aerospace power has performed best when supporting defensively organized ground troops, pitted against easily identifiable opposition forces, in fairly open terrain on the outskirts of small, isolated towns. Close air support has generally been less effective in offensive operations conducted within densely built urban metropolises, where adversary forces have been dispersed in well-fortified defensive positions or intermixed with local civilians. Since the 1970s, developments in command, control, communications, computers, intelligence, surveillance, and reconnaissance (C4ISR) and weapon accuracy and lethality have significantly improved how well advanced aerospace forces have engaged hardened targets in urban areas, often close to friendly troops. Nonetheless, factors such as restrictive rules of engagement (ROE), poor visibility, inadequate air-ground cooperation, insufficient intelligence, potent adversary air defenses, and the opposition's clever use of urban terrain and noncombatants have degraded the effectiveness of CAS, particularly with respect to small, mobile targets, such as snipers, mortars, and rocket-propelled grenade (RPG) launchers. In some cases, as in Panama in 1989, these factors have not altered the overall positive impact of close support. In other cases, as in Grozny in 1994–1995, they have not only posed insurmountable obstacles for CAS but have added to the negative view of the operation as a whole.

Results

With some notable exceptions, neither the Axis nor the Allied powers during World War II had much success in providing close air support in urban areas. As a rule, they employed air forces in the city massively and offensively to "soften up" and demoralize the enemy prior to a major ground assault. In such a role, aerospace power often destroyed countless civilian lives and property, without making a significant military contribution. For example, on July 23–24, 1942, German bombers mounted the equivalent of 2,000 sorties against the city of Stalingrad, killing approximately 40,000 people and, at least initially, causing widespread panic among the Russian population. By blocking roads with the rubble produced by fallen buildings, the preliminary air bombardment hampered the movement of Soviet military forces. But it also assisted the city's defenders by impeding the German ground attack. For their part, Stuka dive-bombers strafed defenseless civilians caught in the open but generally could not provide effective fire support to friendly units attempting to dislodge small groups of Soviet troops from the remains of Stalingrad's municipal buildings and factories.[1]

On the Western Front, aerial bombardment contributed little to Allied assaults on the German-held towns of Cassino, Caen, and Aachen. Used for the first time in a close support role at Cassino, U.S. heavy bombers caused many casualties and undoubtedly demoralized many Germans defending the town. But these benefits were offset somewhat by all the rubble, which impeded the movement of friendly tanks and other vehicles, as well as by the fact that air strikes

---

[1]William Craig mentions one German battalion commander during the initial stage of the Stalingrad battle who, having lost 200 of his men in one day, decided not to pursue a group of Russian snipers into the city's main railway station. Instead, he called for an air strike. The Stukas, however, missed the target and dropped their bombs in the midst of friendly troops. See *Enemy at the Gates*, New York: Reader's Digest Press, 1973, pp. 93–94.

only partially neutralized German machine guns and artillery.[2] In Caen, some friendly units were greatly hindered by rubble in the streets; others elsewhere in town were not affected at all. Nevertheless, the effect on the enemy was clear. After sacrificing up to a quarter of their manpower in the assault on Caen, British infantry units reported almost no evidence of German gun positions, tanks, or German dead in the area targeted by Allied bombs. Instead, what they discovered was a devastated town center and 5,000 dead French civilians.[3] At Aachen, the bombing results were nearly as dismal. Despite the loss of 79 planes and the diversion of precious sorties from the interdiction mission, the U.S. IX Tactical Air Force did nothing to speed the capture of the German border town. Indeed, the German defenders of Aachen managed to hold out for 39 days against an assault force of five U.S. divisions.[4]

However, urban CAS did achieve a small measure of success in World War II. U.S. Army commanders at Cherbourg credited the air support they received from the 9th Air Force in particular with shortening the battle by 48 hr or more. Even so, U.S. Army Air Force intelligence analysts subsequently described the bombing's impact as more psychological than physical. "Flying artillery" had not replaced ground artillery, and, in many cases, Allied ground forces encountered stiff resistance from German strongpoints that had survived pre-assault bombing. Whereas air strikes contributed to the surrender of some German forts, other German garrisons continued to endure for days, giving the Germans time to sabotage the city's valuable harbor facilities.[5]

A more convincing demonstration of Allied air support occurred on the periphery of Bastogne, during the winter of 1944–1945. For a week after the weather finally cleared over the Ardennes on

---

[2]See Headquarters, Mediterranean Allied Air Forces (MAAF), "Air and Ground Lessons from the Battle of Cassino, March 15-27, 1944," Maxwell AFB, Ala.: U.S. Air Force Historical Research Agency, May 4, 1944.

[3]Carlo D'Este, *Decision in Normandy*, New York: E. P. Dutton, 1983, pp. 228–230; and Alexander McKee, *Caen: Anvil of Victory*, New York: St. Martin's Press, 1964, p. 230.

[4]Thomas Alexander Hughes, *Over Lord: General Pete Quesada and the Triumph of Tactical Air Power in World War II*, New York: The Free Press, 1995, pp. 260–261.

[5]"Air Force Operations in Support of Attack on Cherbourg, June 22–30, 1944," Maxwell AFB, Ala.: U.S. Air Force Historical Research Agency.

December 23, P-47 fighter bombers from the XIX Tactical Air Command (TAC) carried out hundreds of precision strikes against German positions all around the besieged town, contributing greatly to the 101st Airborne Division's successful defense of this vital communications center.[6]

Even in the post–WWII period, close air support has usually been less useful when the enemy's main forces have seized the densely populated, built-up areas of a city. For example, because of poor weather and U.S. fear of civilian casualties and damage to the city's historic citadel, CAS was mostly unavailable to American and South Vietnamese troops attempting to retake Hue during the Tet Offensive in 1968. As a result, U.S. Marines could not employ aerospace power against the defending Communist Vietnamese to compensate for their lack of adequate artillery support, thus prolonging the siege and increasing the risk of allied casualties. Still, four CAS missions flown against the southeast wall of Hue's Citadel enabled the Marines to capture an enemy position they had previously failed to seize.[7] During the 1972 Easter Offensive in Vietnam, aerospace power proved essential to the defense of An Loc and Kontum, among other places. At An Loc, U.S. Air Force gunships broke up numerous Communist assaults on the town's perimeter before they were even organized. In addition, B-52 ARC LIGHT strikes destroyed enemy troop formations and eliminated artillery and anti-aircraft positions.[8]

During the Persian Gulf War Battle of Khafji, coalition air forces and Army/Marine helicopters provided effective close support for friendly ground forces around Khafji, especially during efforts to re-

---

[6]10th Armored team commander Colonel William Roberts described the P-47's firepower on December 23 as equivalent to that of two to three divisions. See S. L. A. Marshall, *Bastogne: The First Eight Days*, Maxwell AFB, Ala.: U.S. Air Force Historical Research Agency, 1988, p. 146.

[7]Headquarters, 1st Battalion, 5th Marines, 1st Marine Division, "Command Chronology for Period 1–31 March 1968," Washington, D.C.: Marine Corps Historical Center; and Eric Hammel, *Fire in the Streets: The Battle for Hue Tet 1968*, Pacifica, Calif.: Pacifica Press, 1991, pp. 341–347.

[8]U.S. Air Force, Headquarters PACAF, *The Battle for An Loc, 5 April–26 June 1972: Project CHECO*, Hickam AFB, Hawaii, January 31, 1973, pp. 59–66.

take the town.[9] Much less effective was the close support provided by Russian Fencers and Frogfoots during the Battle for Grozny in 1994–1995. Aside from killing civilians and contributing to the city's ruin, inaccurate Russian air strikes reportedly caused as many casualties to Russian ground troops as did rebel Chechen mortar fire.[10]

Since the 1980s, U.S. close air support assets have participated in a number of other-than-war operations in urbanized locales such as Grenada, Panama, Mogadishu, and Tirana. In most of these, CAS demonstrated its value to American troops and civilians on the ground. But special MOOTW considerations—in particular, those relating to U.S. military and noncombatant casualties—sometimes reduced the effectiveness of CAS. During the 1983 invasion of Grenada, for example, Air Force gunships ensured the safety of 82nd Airborne Division paratroops at Salinas Airfield and protected Navy Sea, Air, Land (SEALs) trapped inside the governor general's compound.[11] For their part, Marine AH-1 Cobras covered the seizure of Pearls airport and supported Army operations on the south end of the island.[12] However, these achievements were offset by the loss of two Marine Cobra helicopters at Fort Frederick and accidental strikes by Navy A-7s on a mental hospital and brigade command post.[13]

---

[9]Rick Atkinson, *Crusade: The Untold Story of the Persian Gulf War*, Boston: Houghton Mifflin, 1993, pp. 198–213; Rebecca Grant, "The Epic Little Battle of Khafji, *Air Force Magazine*, February 1, 1998, pp. 28–34; and Mike Williams; "Battle for Khafji," *Soldier of Fortune*, May 1, 1991 pp. 48–52.

[10]NATO, "Frontal and Army Aviation in the Chechen Conflict," gopher://marvin.nc3a.nato.int/00/secdef/csrc/adv1020%09%09%2B, December 19, 1995 (downloaded November 13, 1998).

[11]Major Mark Adkin, *Urgent Fury: The Battle for Grenada*, New York: Lexington Books, 1989, pp. 183–185, 209–210.

[12]During one action, a flight of two Cobras used 20mm cannon and tube-launched, optically tracked, wire-guided (TOW) missiles to destroy a 90mm recoil-less rifle position, along with the house in which it was located and an adjacent support vehicle. See Timothy A. Jones, *Attack Helicopter Operations in Urban Terrain*, Ft. Leavenworth, Kansas: U.S. Army, Command and General Staff College, December 20, 1996, pp. 7–8; and Lt. Col. Ronald H. Spector, *U.S. Marines in Grenada*, Washington, D.C.: Headquarters, U.S. Marine Corps, History and Museum Division, 1987, pp. 8, 10.

[13]Ronald H. Cole, *Operation Urgent Fury: Grenada*, Washington, D.C.: Office of the Chairman of the Joint Chiefs of Staff, Joint History Office, 1997, pp. 4–5; and Adkin, 1989, pp. 243, 285–287.

In 1989, during Operation Just Cause in Panama, U.S. Army and Marine attack helicopters and Air Force AC-130 gunships provided effective close support in both rural and urban settings.[14] They suppressed anti-aircraft and sniper positions around *La Commandancia*, the main Panamanian Defense Forces (PDF) headquarters complex, as well as the Tocumen and Rio Hato airports, and provided cover for 82nd Airborne Division air assaults against Panama Viejo, Tinajitas, and Fort Cimarron.[15] On the downside, communications difficulties hindered AC-130 support to the SEAL assault on Paitilla Airfield[16] and an AC-130 responding to a request for fire accidentally fired on a friendly ground unit near *La Commandancia.*[17]

During the summer and fall of 1993, U.S. forces participating in the UN mission in Somalia employed helicopter gunships in a number of high-profile CAS operations.[18] On June 5, 1993 one AH-1 Cobra may have saved hundreds of Pakistani and American lives during a road-clearing operation by establishing a cordon of fire around UN troops. On September 10, Cobra gunships fired on Somali gunmen swarming around UN forces. Most famously, on October 3–4, 1993, four AH-6 Little Birds prevented U.S. Ranger Task Force troops pinned

---

[14]In its lessons-learned volume, the U.S. Army observed that "all major units involved in *Just Cause* used the AC-130. It provided precise direct fire, night surveillance and navigation assistance. . . . The AC-130 is an excellent fire support system. Precision fire control and accurate weapons systems fit well within restrictive ROE and reduction of collateral damage." See *Operation Just Cause Lessons Learned: Volume II: Operations*, Ft. Leavenworth, Kansas: Center for Army Lessons Learned, October 1990, p. II-8.

[15]Later, Cobra and Apache attack helicopters supported ground units involved in clearing operations in Panama City and Colon. Jones, 1996, pp. 9–11.

[16]Malcolm McConnell, *Just Cause: The Real Story of America's High-Tech Invasion of Panama*, New York: St. Martin's Press, 1991, p. 66.

[17]Soldiers belonging to the 2nd platoon of D Company, 6th Regiment, 5th Infantry Division claim that half of their members were wounded by Spectre cannon fire as they attempted to breach a fence surrounding the PDF headquarters. See Thomas Donnelly, Margaret Roth, and Caleb Baker, *Operation Just Cause: The Storming of Panama*, New York: Lexington Books, 1991, pp. 150–152.

[18]John R. Murphy, "Memories of Somalia," *Marine Corps Gazette*, April 1998, pp. 22–23; Jones, 1996, p. 12; and Colonel William C. David, "The United States in Somalia: The Limits of Power," *Viewpoints*, Vol. 95, No. 6, June 1995, p. 9.

down in the vicinity of Mogadishu's Bakara Market from being overrun by supporters of Mohammed Farah Aideed.[19]

In the only serious incident of the Tirana NEO in 1997, U.S. Marine Cobras used cannon and rocket fire to take out two threatening Albanian air defenders, one equipped with an SA-7 launcher and the other manning a 12.7mm machine gun, on a ridgeline near the U.S. Embassy compound. From that point on, the Marines had no problems with Albanian gunmen attempting to disrupt airlift operations.[20]

Effectiveness Factors

The following is a list of factors that have contributed to the effectiveness (or ineffectiveness) of the preceding urban CAS operations. They are grouped into performance categories: weapons and equipment, command and control, rules of engagement, intelligence, tactics and training, logistics, ground-force cooperation, opposition countermeasures, atmospheric and light conditions, and geography and terrain.

In most cases, no one factor or performance category has been responsible for the overall effectiveness of a given urban operation. Nonetheless, certain conclusions can be drawn from the available historical evidence:

- Technological advances since the 1970s have generally enhanced the performance of CAS-related weapons and command and control systems in urban environments.

- Strict or complex ROE in effect since the Vietnam War have sometimes offset technological advances in urban CAS.

---

[19]The most comprehensive account of the Ranger Task Force engagement on "Bloody Sunday" is Mark Bowden's *Black Hawk Down: A Story of Modern War*, New York: Atlantic Monthly Press, 1999. See also Jonathan Stevenson, *Losing Mogadishu: Testing U.S. Policy in Somalia*, Annapolis, Md.: Naval Institute Press, 1995.

[20]LtCol. Jon T. Harwick, CO HMM 365, *Operation Silver Wake*, Oral History Interview, Washington, D.C.: Marine Corps Historical Center, Marine Corps Oral History Program, June 12, 1997, p. 14. See also Jon R. Anderson, "Rescue 911: On the Ground with Marines in Albania," *Navy Times*, March 31, 1997, p. 13.

- Passive and active countermeasures employed by opposing forces have degraded urban CAS performance from WWII on, despite advances in weapons and C2.

- Adverse weather and urban terrain have remained significant obstacles to close air support; however, improvements in precision guided munitions and aircraft navigation and targeting have reduced their impact to some extent.

Weapons/Equipment:

- At Cassino and Aachen, U.S. air forces lacked heavy, delayed-action bombs that could reach into cellars and penetrate concrete emplacements occupied by German troops.[21]

- Napalm proved to be the most effective ordnance at Hue because of the enemy's dug-in positions, proximity to friendly ground troops, and cover of well-constructed cement buildings. Useful also was the delayed-action Snakeye bomb, which could be released at low altitude without knocking down the aircraft that dropped it.[22]

- Employed for the first time in an urban CAS role at An Loc,[23] the AC-130's 105mm howitzer and PAVE AEGIS targeting system proved invaluable in that city's close-quarters fighting. Provided with a map of the city, Spectre crews were able to follow detailed instructions from ground controllers as well as break up enemy assaults and destroy buildings close to friendly troops.[24]

- The Grenada operation highlighted certain deficiencies in the tools that the United States was then employing for urban CAS, including the AH-1 Cobra's lack of armor protection and the inability of the Navy's A-6 and A-7 aircraft to accurately acquire and hit targets in a built-up area.

---

[21]Headquarters, MAAF, 1944, p. 6; and Hughes, 1995, p. 63.

[22]Hammel, 1991, p. 59.

[23]Prior to An Loc, the AC-130 Spectre's mission in Indochina had been primarily night interdiction and armed reconnaissance, with less support of troops in contact. U.S. Air Force, Headquarters PACAF, 1973, pp. 59–60.

[24]U.S. Air Force, Headquarters PACAF, 1973, pp. 59–60.

- In Panama, the Hellfire missile launched from the AH-64 Apache helicopter proved ineffective against the steel-reinforced walls of *La Commandancia.*[25]

- At Khafji, the nighttime navigation and targeting capability of American close support aircraft, both fixed-wing and rotary-wing, was key to preventing Iraqi armored forces from entering the city en masse.[26]

- In Somalia, the small and agile AH-6 gunship proved itself an ideal close support platform in Mogadishu's densely populated urban environment.[27]

- When bombing through clouds, although equipped with a limited number of PGMs, Russian Su-24 Fencers at Grozny were still not accurate enough to avoid considerable collateral damage or fratricide.[28] Lacking guided missiles and bombs, the Su-25 Frogfoot's delivery was generally less accurate than the Fencer's.[29]

Command and Control:

- Under heavy pressure from General Dwight Eisenhower and Prime Minister Winston Churchill to produce results, Field

---

[25]Donnelly et al., 1991, pp. 119, 157; and McConnell, 1991, p. 66.

[26]Williams, 1991, p. 50.

[27]Bowden, 1999, p. 340. Throughout the night of October 3–4, Little Bird helicopters prevented Somali fighters from overrunning isolated U.S. positions, laying down fire as close as 20 feet from friendly troops. By contrast, Cobra pilots had difficulties identifying targets designated by ground controllers. See CPT Drew R. Meyerowich, Commander, Alpha Company, 2/14 Infantry, 10th Mountain Division, Interview, Washington, D.C.: U.S. Army Center of Military History, April 18, 1994, pp. 38, 46–48.

[28]NATO, 1995, p. 2. With better weather and a more extensive network of ground controllers, LGB-equipped Su-24s performed much better against the smaller Chechen strongholds of Argun, Gudermes, and Shali during spring 1995. See Sean J.A. Edwards, "Mars Unmasked: The Changing Face of Urban Operations," unpublished RAND research, p. 73.

[29]Still, the most notable Russian air force success during the first battle of Grozny was the bombing of the Presidential Palace on January 17, 1995, by seven Su-25s. Two of the 3000-lb concrete-piercing bombs penetrated the palace from top to bottom, leaving the Chechens inside in shock and probably causing the building to be evacuated. See Benjamin S. Lambeth, *Russia's Air Power in Crisis,* Washington, D.C.: Smithsonian Institution Press, 1999, p. 125.

Marshal Bernard Montgomery hastily requested heavy bombers for use against Caen. Although there were a number of worthwhile German targets in the city, none was apparently included in the target area selected by the army.[30]

- The entire air operation during the Battle of Bastogne was carefully systematized and supervised. After directing flights to the town, the air force ground controller brought fighter-bombers straight over the target, eliminating the need to search. Planes were then ordered to reconnoiter Bastogne's perimeter, providing targets for succeeding flights.[31]

- At An Loc, the air commander established a so-called King Forward Air Controller (FAC) to sort out the myriad aircraft on the scene associated with different commands, services, and countries. Furthermore, the diversion of B-52 strikes to higher-priority targets became standard operating procedure at An Loc.[32]

- Early CAS operations in Grenada exhibited poor command, control, and communications: Attacks by Navy fighters were not well coordinated with Rangers and Special Forces on the ground; furthermore, Marine Cobra pilots had difficulty making radio contact with Air Force C-130s or Army ground units.[33]

- During the gun battle between the Navy SEALs and the PDF at Paitilla Airport in Panama, the Air Force Combat Control Team (CCT) was unable to make radio contact with the AC-130 Spectre overhead that had been assigned to provide close air support.[34]

- In Khafji, air liaison officers working with Marine and Arab units and airborne observer-controllers were generally able to direct

---

[30]D'Este, 1983 p. 311.

[31]Marshall, 1988, pp. 145–146.

[32]Over 90 percent of B-52 missions were changed at the last minute by the local FAC. U.S. Air Force, Headquarters PACAF, 1973, pp. 18–19, 64–65.

[33]Close air support was also complicated by Marine pilots and Army Rangers using different maps. Cole, 1997, pp. 65–66; and Adkin, 1989, pp. 217–218.

[34]McConnell, 1991, p. 63.

air assets against attacking Iraqi units in the early hours of the battle.[35]

- Soldiers of the 10th Mountain Division quick-reaction force (QRF) in Somalia were equipped with infrared strobes, which effectively enabled attack-helicopter pilots to distinguish them from adversaries during periods of close-in combat at night.[36]

- As a consequence of an overly complex air control system and the threat posed by Chechen gunfire to air controllers, Russian air forces operated virtually independently of ground units during the Battle of Grozny. This lack of communication resulted in many fratricides.[37]

Rules of Engagement:

- During the Battle of Hue, allied concern for the safety of the city's remaining civilian residents initially resulted in restrictive ROE, which, along with the bad weather, countered U.S. and Army of the Republic of Vietnam (ARVN) advantages in aerial fire support.[38]

- In Grenada, the likelihood of civilian casualties led to the selection of the AH-1 Cobra (over Naval gunfire or carrier aircraft) for the fateful attack on Fort Frederick, during which gunship pilots flew a fixed course for a risky length of time. As a result, one Cobra was shot down and another was lost while providing fire support for the rescue of the downed pilots. Both Cobras were destroyed, and three crewmembers died.[39]

- The ROE in Beirut permitted U.S. forces to shoot only in self-defense, and then only if the target could be positively identified and accurately engaged. Consequently, very few Marine Cobras were even allowed to fly over the beach; and, although transport

---

[35]Williams, 1991, p. 50.

[36]Meyerowich, 1994, p. 47.

[37]NATO, 1995, p.5.

[38]MAJ John C. Latimer, "Considerations for Operations on Urban Terrain by Light Forces," Master's thesis, Ft. Leavenworth, Kansas: U.S. Army Command and General Staff College, 1985, p. 108.

[39]Adkin, 1989, pp. 242–245.

helicopters were armed with .50-caliber machine guns, their gunners never fired a round.[40]

- Tactical air operations in Panama were often tightly controlled by ground force commanders. When civilians were present, the employment of AC-130 tube- or rocket-launched weapons was prohibited without the permission of a ground commander with at least the rank of lieutenant colonel. Close air support in civilian areas required approval from at least division level. The commander of Operation Just Cause, Lt Gen Carl Stiner, authorized air strikes for fighter aircraft.[41]

- Because of ROE restrictions on the use of mortars, Cobra attack helicopters were often the only fire support available to the United Nations Task Force commander in Somalia. To help reduce collateral damage, the Cobras' 20mm cannons were fitted with an AIM-1 laser designator, permitting the gunner to score first-round hits at night, when the laser was visible to night-vision goggles.[42]

- While they were in effect, Russian ROE in Chechnya restricted the use of air-to-ground munitions in civilian areas. However, these ROE were eventually violated because of the limited supply of precision-guided weapons, poor weather, and a lack of training. Heavy civilian casualties resulted.[43]

- The local Marine commander's insistence that ROE be loosened to permit the use of Cobra gunships at Tirana proved essential to

---

[40]Although these restrictions precluded close air support for troops on the ground, they may have limited the number of aircraft targeted by the Lebanese militia. LtCol. Larry Medlin, CO, HMM-162, November 20, 1983, pp. 13–14; and LtCol. Amos R. Granville, CO, HMM-261, Oral History Interview, Washington, D.C.: Marine Corps Historical Center, Marine Corps Oral History Program, May 22, 1984, pp. 7–8.

[41]Jennifer Morrison Taw, *Operation Just Cause: Lessons for Operations Other Than War*, Santa Monica, Calif.: RAND, MR-569-A, 1996, p. 24.

[42]Jones, 1996, p. 12.

[43]Nevertheless, most of the 10,000 to 40,000 civilian deaths in Grozny by August 1995 were caused by artillery fire, not aerospace power. See "The Casualties of Chechnya," *The New York Times*, August 10, 1995, p. 18.

protecting transport helicopters involved in evacuating the U.S. Embassy compound there in 1997.[44]

Intelligence:

- In Grenada, U.S. intelligence failed to anticipate the degree of initial resistance by Cuban advisers. Anticipating a short military intervention, planners were forced to put together a Tactical Air Control system on the fly.[45]

- The complexity of the political-military situation during the U.S. peacekeeping operation in Beirut in the early 1980s made it extremely difficult to know whether any one position would be targeted and, if it was, who lay behind the attack.[46]

Tactics and Training:

- The tendency of XV Air Force bombers at Cassino to drop their munitions from too high an altitude, given the slight anti-aircraft threat and good visibility, meant that bombing accuracy was worse than it might otherwise have been.[47]

- Low-altitude dive-bombing and strafing, in close coordination with troops on the ground, proved effective in suppressing German defenses at Cherbourg.[48]

- In an attempt to avoid hitting friendly troops through "backsliding," many aircraft at Caen ended up bombing ahead of the target, thus adding to the devastation of the city center while causing little harm to the enemy.[49]

---

[44]LtCol. Dan E. Cushing, X0-HMM-365, *Operation Silver Wake*, Oral History Interview, Washington, D.C.: Marine Corps Historical Center, Marine Corps Oral History Program, June 12, 1997, p. 13.

[45]Adkin, 1989, p. 140.

[46]Hammel, 1991, p. 196.

[47]Headquarters, MAAF, 1944, p. 10.

[48]"Air Force Operations in Support of Attack on Cherbourg," 1944, p. 5.

[49]*Backsliding* refers to the tendency of a bomber pilot as he neared the drop zone to drop his load as soon as was acceptable. When the majority of bombs were dropped on the near edge of the zone, the zone itself would begin to slide back from the target. Because "backsliding" at Caen would have meant bombing friendly ground forces,

- In Grenada, AC-130s were not earmarked specifically to support Navy SEAL operations in the vicinity of St. George's. This failure contributed to U.S. and prisoner casualties resulting from People's Revolutionary Army armored personnel carrier (APC) and mortar fire.[50]

- In attempting to minimize casualties and collateral damage in the crowded residential areas of Quarry Heights and Albrook Air Station, U.S. forces in Panama successfully employed a "graduated response" technique. This technique involved loud-speaker appeals to surrender, combined with a nearby firepower demonstration by AC-130 gunships, threatening imminent destruction unless the adversary gave up.[51]

- The AC-130 fratricide incident in Panama might have been avoided had AC-130s exercised more frequently with heavy forces in urban environments.[52]

- Although subsequently provided, air cover (and armored support) to the 10th Mountain Division's Quick Reaction Force was lacking during the evening of October 3, 1993. This lack contributed to the unit's initial failure to break through a Somali ambush and enact a timely rescue of the Rangers pinned down in the vicinity of Mogadishu's Bakara Market.[53]

Logistics:

- Because the movement of fighter groups had not kept pace with the Allied ground advance, U.S. fighter-bombers were forced to operate from airfields over 100 miles from the front lines during the Battle of Aachen. The resulting logistics difficulties cost IX TAC almost one-third of its striking power.[54]

---

many airmen overcompensated and actually dropped on the far side of their target. McKee, 1964, pp. 228–230.

[50]Adkin, 1989, p. 183.

[51]In both cases, the PDF soldiers either surrendered or fled. See Donnelly et al., 1991, pp. 143–144, 153.

[52]Donnelly et al., 1991, p. 405.

[53]Col. David Hackworth, "Rangers Ambushed in Somalia," *Soldier of Fortune,* Vol. 19, January 1, 1994, pp. 93–94.

[54]Hughes, 1995, p. 262.

- When military operations commenced in Chechnya, supplies of food, fuel, ammunition, and spare parts amounted to 50 percent of those required. The shortage in material resources compromised the ability of Russian aircrews to operate in adverse weather and to employ their weapons effectively.[55]

Ground-Force Cooperation:

- At Stalingrad, Luftwaffe General Wolfram von Richthofen berated the German commander, von Paulus, for not taking better advantage of the suppressive power provided by von Richthofen's Stukas and Junker bombers so that ground assaults could be launched into the city.[56]

- Likewise, U.S. Army Air Force's reports on the Battle of Cassino are critical of the infantry for not advancing quickly enough under barrage, as well as for relying too heavily on bombardment to neutralize German defensive positions.[57]

- In the defensive battle of Bastogne, U.S. ground units and Army Air Force fighter-bombers performed as part of a well-honed combined-arms team, breaking up numerous German armored attacks.[58]

- Following the Grenada invasion, U.S. infantrymen were criticized for their failure to advance in the face of light opposition without overwhelming air and artillery support.[59]

Opposition Countermeasures:

- At Stalingrad, Soviet commander Lieutenant General Vasily Ivanovich Chuikov's tactic of employing squad-sized "storm groups" in strategic buildings, a city-hugging tactic, hampered

---

[55]NATO, 1995, p. 1.

[56]Craig, 1973, p. 133.

[57]Headquarters, MAAF, 1944, p. 8.

[58]See Michael D. Doubler, *Closing with the Enemy: How GIs Fought the War in Europe, 1944–1945*, Lawrence, Kansas: University Press of Kansas, 1994, pp. 220–221.

[59]Adkin, 1989, p. 339.

the Germans' ability to coordinate artillery and air support because doing so risked engaging their own troops.[60]

- By contrast, the German practices at Bastogne of keeping their armored and support vehicles on the roads and, at least initially, refraining from using their anti-aircraft guns increased the effectiveness of American close air support.[61]

- During the Battle of Hue, Communist Vietnamese anti-aircraft fire drove helicopter gunships from the city and made conditions extremely difficult for airborne observers.[62]

- In Grenada, the enemy had no air force or radar-controlled air defenses with which to challenge the attack helicopters and those AC-130 gunships providing close support to invading U.S. ground forces.[63]

- Most of the anti-aircraft fire directed at Marine helicopters during the Beirut peacekeeping mission was limited to small arms and some RPGs. However, even that minimal opposition was sufficient to severely restrict gunship operations over the city.[64]

- In Panama, the opposition's lack of effective air defenses contributed greatly to the success of aerospace power.[65] However, the PDF's use of human shields limited the ability of attack helicopters to return fire during the 82nd Airborne operations at Panama Viejo and Tinajitas barracks. At Panama Viejo, PDF soldiers ducked into a crowd of civilians after firing. At Tinajitas, they fired from civilian buildings near the objective.[66]

---

[60]Colonel Michael Dewar, *War in the Streets: The Story of Urban Combat from Calais to Khafji*, New York: David & Charles, 1992, p. 21; and Craig, 1973, p. 91.

[61]Marshall, 1988, p. 144.

[62]Jones, 1996, p. 5.

[63]Adkin, 1989, p. 197.

[64]Maj. Dick Gallagher, USMC OpsO, HMM-261, Oral History Interview, Washington, D.C.: Marine Corps Historical Center, Marine Corps Oral History Program, May 22, 1984, pp. 12–13.

[65]Even so, 30 percent of the Special Operations aircraft in Panama were damaged or shot down, including the AH-6 carrying American civilian Kurt Muse; Taw, 1996, p. 21.

[66]Unlike the Apaches, Cobra gunships did engage some PDF positions at Tinajitas with rockets and cannon. See Jones, 1996, p. 10; and Donnelly et al., 1991, pp. 222–223.

- In Mogadishu, the distinction between combatants and non-combatants became very murky. Both when they moved toward the Task Force Ranger helicopter crash sites and when they fired on U.S. personnel, Somali fighters, who wore no uniforms or distinctive clothing, hid behind mobs of unarmed men, women, and children. Further complicating matters, much of Mogadishu's population in the Bakara Market area and along relief-convoy routes rose up at this time against U.S. forces.

- Chechen rebels in Grozny countered Russian air superiority by deploying their tanks and guns in residential areas; attacking from hospitals, schools, and apartment blocks; and even breaking into Russian radio transmissions and directing Russian aircraft over the Russian's own troops.[67] Furthermore, the Chechen air defenses—which included SAMs (SA-13s and SA-16s) and radar-controlled AAA, in addition to heavy machine guns and RPGs—proved highly lethal to helicopters. As a result, the Russians used helicopters mostly for noncombat missions.[68]

Atmospheric and Light Conditions:

- During the first two days of the Bastogne battle, fog served as a protective screen for the American defenders and created confusion in the German ranks. However, the fog prevented Allied aircraft from providing support to engaged U.S. forces.[69]

- At Hue, however, consistently low cloud ceilings, combined with restrictive ROE, prevented most close-support operations for three weeks. When aircraft were permitted to fly, they often did so at low altitudes, making it impossible to track targets.[70]

---

[67] NATO, 1995, pp. 2, 6.

[68] By May 1996, a total of 14 Russian helicopters had been lost and 30 damaged. Several more were shot down during the final battle for Grozny in 1996. See Anatol Lievan, *Chechnya: Tombstone of Russian Power*, New Haven and London: Yale University Press, 1998, p. 278.

[69] Marshall (1988, pp. 140, 145) notes that the winter weather also helped the Allies by allowing aircraft to identify enemy armored positions in the forest by their tracks in the snow.

[70] Hammel, 1991, pp. 58–59.

- Before the beginning of Operation Just Cause, a North Carolina ice storm delayed the departure of the 82nd Airborne Division. This delay caused airborne assault operations in the vicinity of Panama City that had been scheduled to be conducted under cover of darkness to be done in daylight, putting paratroopers at greater risk from adversary ground fire.[71]

- Following the Panama invasion, AC-130 pilots indicated that smoke and fire may have obscured their targeting systems during the battle for *La Commandancia,* possibly contributing to the friendly-fire incident there.[72] It also appears that ambient light in Panama City washed out the reflections from identification tape on U.S. vehicles, making them look like Panamanian armored vehicles.[73]

- During the initial assault on Grozny, poor weather—blowing snow, ice, and low cloud ceilings—ruled out visual bombing, as well as the use of electro-optical or laser-guided weapons. As a result, Russian Su-24 Fencers were forced to radar-bomb from medium altitude, which led to inaccurate deliveries and many Russian losses to friendly fire.[74] In addition, Russian helicopters were grounded for most of the month of February 1995 because they lacked all-weather capabilities.[75]

Geography and Terrain:

- At Cassino, the Germans made good use of the town's cellars and existing tunnels. Furthermore, their position atop Monte Cassino gave them unobstructed observation of Allied movements in the town, no matter how heavy the artillery fire or bombing.[76]

---

[71]The weather also forced a change of plan with regard to the 82nd's arrival at Torrijos International Airport. Rather than landing on the airport runway, the transport aircraft dumped the paratroopers from the air in three waves. In the process, paratroopers became intermixed with Rangers involved in clearing operations on the ground, some of whom were still engaged in minor firefights with the PDF. Fortunately, the 82nd did not suffer any casualties during the drop. Donnelly et al., 1991, pp. 200–203.

[72]Donnelly et al., 1991, p. 152.

[73]Interview with USAF pilot who flew on *La Commandancia* mission.

[74]Lambeth, 1999, p. 124.

[75]Edwards, unpublished research, p.74.

[76]Headquarters, MAAF, 1944, pp. 6, 8.

- An integral part of the German Siegfried Line, Aachen had numerous strongpoints on each town flank and bunkers built of solid concrete that could stand up under a direct hit with a 500-lb bomb.[77]

- In Beirut, Marine helicopter pilots claimed that they flew too fast to identify or hit targets in densely populated neighborhoods, even with the assistance of airborne FACs and gyro-stabilized binoculars.[78]

- U.S. forces stationed in Panama had the good fortune of fighting over familiar terrain, thus reducing the psychological stress and uncertainty inherent in combat.

- With its open layout—of mostly low, modern buildings, practically devoid of civilian inhabitants—Khafji was a particularly favorable venue for the limited urban CAS conducted during that battle.[79]

- On the one hand, the generally low-rise environment around Bakara Market, and the fact that most of the fighting was conducted in the open or from just inside buildings, provided relatively good fields of fire for attack helicopters flying close support operations on "Bloody Sunday." On the other hand, the low-rise environment and the neighborhood's narrow streets may have increased the danger to helicopters from RPG launchers.[80]

LOGISTICS SUPPORT

From Leningrad to Sarajevo, aerial resupply and transport have played an important role in major urban operations. However, the inherent vulnerability of most logistics support aircraft has limited their employment and effectiveness in highly contested urban environments, particularly when the opposition has possessed significant air defenses. During WWII, aerial resupply operations were con-

---

[77]Hughes, 1995, p. 258.

[78]Medlin, 1983, pp. 14–15.

[79]Dewar, 1992, p. 82; Williams, 1991, pp. 48–49.

[80]For a drawing and description of the Mogadishu battle site, see Bowden, 1999, pp. 3, 12.

strained by adverse weather, the unavailability of appropriate aircraft, long distances between air bases and the landing (or drop) zones, insufficient intelligence on the locations of friendly and enemy forces, and the lack of a precise airdrop-delivery mechanism.

Since the 1970s, technological advances have enabled airdrops to be made with great accuracy from high altitudes, at night and in poor weather. Even so, ensuring that the right people get the supplies has remained a problem. Furthermore, even well-executed airdrop operations cannot match the volume of cargo that can be moved through airport, sea, or ground alternatives.

Since WWII, the effectiveness of air-transport operations in contested urban areas has relied on one element: surprise. Prior to the 1960s, getting ground forces to the objective as quickly as possible— before the opposition had time to mount a coordinated defense— required either that they be air-dropped or air-landed by transport plane or glider, preferably close to the target. To be effective, such an operation usually needed good weather, an exceptionally well-trained infantry force, a relatively weak opposition, and a large amount of luck.

Since the Vietnam War, helicopters have performed most troop-transport missions within urban areas. Their small size and maneuverability relative to transport planes—and, more recently, their aerial-refueling and nighttime capabilities—have enabled helicopters to drop off and pick up hundreds of individuals in fairly close urban terrain and to transport them safely over considerable distances. These characteristics have made helicopters especially useful in urban-related NEOs. Nonetheless, recent U.S. engagements in Somalia have demonstrated that even armored transport helicopters can be brought down by relatively unsophisticated weapons such as RPGs, making their employment problematic in nonpermissive urban situations.

Results

Most of the major WWII air resupply operations conducted in urban areas ended in failure. During the siege of Leningrad, the early Soviet attempt to bring in emergency supplies by air transport fell far short of meeting the city's needs. The Soviet Air Force was able to fly in a mere 3,357 tons of food during the last two and a half months of 1941, even though the city's requirement for flour alone was 1,000 tons a day. Partly as a consequence, Leningrad seldom had more than one or two days of food in reserve before the Soviet government organized the massive resupply operations across Lake Ladoga.[81]

During the winter of 1942, the German Luftwaffe launched an impressive number of airlift sorties in an attempt to replenish the stores of the beleaguered Sixth Army at Stalingrad. However, many of the aircraft had to be diverted, and only one-sixth of the supplies needed ever reached the troops.[82] Mounted with minimal Soviet support and at a great cost in aircraft and aircrews, the Allied airdrop operation during the Polish uprising in Warsaw failed to redress the balance of forces in the city. Indeed, many of the containers fell into enemy hands.[83] On the plus side, the huge aerial resupply during the second week of the Battle of Bastogne helped ensure the town's successful defense, primarily by refilling the 101st Airborne Division's dwindling stocks of artillery ammunition and medicine.[84]

Since WWII, the United States and its allies have conducted several moderately effective air resupply operations in embattled urban areas. After Communist anti-aircraft fire halted low-level transport flights into An Loc, U.S. experts improved high-level radar delivery techniques and also successfully tested a new Adverse Weather Aerial Delivery System (AWADS). As a result, allied aircraft were able to meet An Loc's requirements of about 28 short tons per day, breaking the Communist siege.[85] During the recent war in Bosnia, the United States and its NATO allies carried out the longest humanitarian airlift in history—three and a half years—primarily to help sustain the

---

[81]Leon Gouré, *The Siege of Leningrad*, Stanford, Calif.: Stanford University Press, 1962, p. 153.

[82]Craig, 1973, p. 234. See also Joel S. A. Hayward, "Stalingrad: An Examination of Hitler's Decision to Airlift," *Airpower Journal*, Vol. 11, Spring 1997, pp. 21–37.

[83]Polish volunteers lost 30 percent of their aircraft during initial Warsaw resupply operations. Jozef Garlinski, *Poland in the Second World War*, New York: Hippocrene Books, 1985, pp. 285–288.

[84]Marshall, 1988, pp. 138–139.

[85]See U.S. Air Force, Headquarters PACAF, *Airlift to Besieged Areas, 7 April–31 August 72: Project CHECO Southeast Asia Report*, Maxwell AFB, Ala.: U.S. Air Force Historical Research Agency, pp. 1–48.

Muslim population of Sarajevo and eastern Bosnia.[86] The impact of this massive endeavor was most obvious in the Bosnian capital. For many months during the Serbian siege of Sarajevo, 85 percent of the aid reaching the city arrived via airlift, saving tens of thousands of residents from starvation.[87]

More difficult to assess, however, are the airdrops that occurred over eastern Bosnia. Although accurately delivered in most cases, supplies were mistakenly dropped into Serbian hands at Cerska and Konjevic. These airdrops certainly helped alleviate the suffering caused by the war, but they did not achieve the more-ambitious goal of breaking sieges of towns such as Srebrenica.

The U.S. and its allies have had mixed results with regard to airborne and air assault operations in urban areas. During World War II's Operation Market-Garden in Holland, the British 1st Airborne Division suffered a disastrous defeat at Arnhem Bridge, losing all but 17 members of its original 509-man assault force.[88] By contrast, the U.S. 82nd Airborne Division not only captured all four bridges in the vicinity of Nijmegan, but also defeated every enemy attempt to re-take them.[89]

The two airborne rescue operations in the Congo, jointly conducted by the United States and Belgium in November 1964, were only partly successful. Although these troops achieved their primary purpose of rescuing a large number of hostages and refugees from rebel forces, in the Stanleyville mission, Congolese rebels managed to kill 27 Americans and Europeans before the rescue force arrived at the Victoria Residence Hotel; at least another 50 hostages were soon executed elsewhere, probably in revenge for DRAGON ROUGE. In addi-

---

[86]By January 1996, 12,895 sorties had been flown as part of Operation Provide Promise, bringing in more than 160,000 metric tons of food, medicine, and other relief supplies. Master Sgt. Louis A. Arana-Barradas, "A 'Promise' of Peace: Sarajevo Humanitarian Airlift Ends, New Hope Begins," *Airman*, March 1996, p. 43.

[87]Arana-Barradas, 1996, p. 43.

[88]Eric Niderost, "Gallant Defense at Arnhem Bridge," *World War II*, Vol. 11, January 1, 1997, p. 81.

[89]Although follow-on ground forces failed to cross the Waal River in strength, the 82nd's presence on the south bank provided a springboard for the final Anglo-American offensive that began in spring 1945. James M. Gavin, *On to Berlin: Battles of an Airborne Commander, 1943–1946*, New York: Viking Press, 1978, pp. 190–191.

tion, unexpectedly strong international opposition to the U.S.–Belgian intervention led U.S. leaders to cancel rescue operations planned for the towns of Bunia and Watsa.[90]

More recently, the Ranger assault operation in Mogadishu against Mohammed Farah Aideed's clan stronghold succeeded in its assigned task of arresting approximately a dozen of the warlord's top lieutenants. Nevertheless, the unanticipated downing of two UH-60 Blackhawk transport helicopters by members of Aideed's militia contributed to the deaths of 18 American soldiers, as well as to the Clinton administration's decision to pull U.S. troops out of Somalia.

Over the past several decades, NEOs in various corners of the world have tested U.S. airlift capabilities. Although no U.S. transport aircraft suffered combat damage during the evacuation from Saigon in 1975, poor contingency planning resulted in thousands of mostly Vietnamese evacuees being left to face the victorious Communist army.[91] More-recent NEOs have gone considerably better. In the early 1990s, the 22d Marine Expeditionary Unit (MEU) helicopters evacuated more than 2,400 people from the Liberian capital of Monrovia. Despite constant fighting in the vicinity, the 7-month-long mission was completed without casualties.[92] In January 1991, Marine CH-53E and CH-47 helicopters flew over 400 nautical miles at night to rescue 281 persons from 30 countries in war-torn Mogadishu. Although two Marines were nearly left behind, the Mogadishu operation was conducted without loss of life or injury. Furthermore, all participating U.S. military forces were back on

---

[90]Fred E. Wagoner, *Dragon Rouge: The Rescue of Hostages in the Congo*, Washington, D.C.: National Defense University Research Directorate, 1980, pp. 197–199. See also Major Thomas P. Odom, *Dragon Operations: Hostage Rescues in the Congo, 1964–1965*, Fort Leavenworth, Kansas: Combat Studies Institute, U.S. Army Command and General Staff College, 1988, pp. 153–160.

[91]U.S. Air Force, Office of PACAF History, *The Fall and Evacuation of South Vietnam*, Maxwell AFB, Ala.: U.S. Air Force Historical Research Agency, April 30, 1978, pp. 161–163. See also Oliver Todd, *Cruel April: The Fall of Saigon*, trans. Stephen Becker, New York and London: W. W. Norton, 1990, pp. 346–369.

[92]LtCol. Glen R. Sachtleben, "Operation SHARP EDGE: The Corps' MEU (SOC) Program in Action," *Marine Corps Gazette*, Vol. 75, November 1991, pp. 77–86. See also LtCol. T. W. Parker, "Operation Sharp Edge," *U.S. Naval Institute Proceedings*, Vol. 117, May 1991, pp. 103–106.

station prior to the commencement of Desert Storm operations against Iraq.[93]

Effectiveness Factors

The following is a list of factors that have contributed to the effectiveness (or ineffectiveness) of the preceding urban air logistics support operations. They are grouped into performance categories: equipment, command and control, political factors, intelligence, tactics and training, opposition countermeasures, atmospheric and light conditions, and geography and terrain. As with close air support, no one factor or performance category dominates. Nevertheless, the historical evidence appears to support the following conclusions:

- Aerial resupply technology has performed quite well since the Vietnam War; however, the vulnerability of transport aircraft remains a significant problem in contested urban areas.

- Political factors are important in special operations and NEOs that involve air transport.

- Superior tactics and training have played a large role in the success of air transport operations in urban areas.

- The presence or absence of significant adversary air defenses has remained a critical factor in urban air logistics operations since World War II.

- Although they have often created difficulties, adverse atmospheric and geographic conditions have generally not had a decisive effect on urban air logistics operations.

---

[93]Adam B. Siegel, *Eastern Exit: The Noncombatant Evacuation Operation (NEO) from Mogadishu, Somalia, in January 1991*, Alexandria, Va.: Center for Naval Analyses, 1992, pp. 41–42.

Equipment:

- That only 20 or so transport aircraft (out of 64) were operational at any time substantially diminished the ability of the Soviet military to supply the city of Leningrad by air.[94]

- The unreliability and vulnerability of the German Ju-52 transports contributed to the failure of the Stalingrad airlift. Although these aircraft were later supplemented by He-111 bombers, weather problems prevented the latter from being employed extensively.[95]

- The airborne portion of Operation Market-Garden was hindered by a shortage of troop-carriers and the inability of American aircrews to operate with assured accuracy at night.[96]

- For daytime, high-altitude airdrops during the Battle of An Loc, the U.S. Air Force relied on the Ground Radar Aerial Delivery System (GRADS) to guide C-130 transport planes to a Computer Aerial Release Point (CARP) aligned with the drop zone inside the town.[97] In part to counter the threat to the resupply effort posed by SA-7s, the USAF began using the Adverse Weather Aerial Delivery System for high-level drops in conditions of low visibility or even total darkness.[98]

- All but invulnerable to small-arms fire, the armored Blackhawk helicopter, used to transport troops during Ranger Task Force operations in Mogadishu, proved unexpectedly vulnerable to

---

[94]Gouré, 1962, p. 109.

[95]Craig, 1973, pp. 220–221, 226.

[96]However, the U.S. 82nd and 101st Divisions took better advantage of their limited transport resources than did the British 1st Division. Thus, the Americans ended up with three brigades in action on their first day of battle, whereas the British had less than two brigades immediately available for combat. See Maurice Tugwell, *Arnhem: A Case Study*, London: Thornton Cox, 1975, pp. 26–27.

[97]Initially, however, parachute malfunctions and improper rigging caused most of the supply bundles to drift outside the narrow confines of the drop zone. U.S. Air Force, Headquarters PACAF, 1972, p. 10.

[98]U.S. Air Force, Headquarters PACAF, 1973, pp. 58–59.

Somali RPG fire. During the October 3–4 firefight, two helicopters were shot down and three were damaged and forced to retire.[99]

- In Bosnia, Kevlar armor was added to prevent small-arms rounds from penetrating the flight decks of allied transport aircraft.[100] In addition, most countries involved in the Sarajevo airlift used C-130s, which meant that little materiel could be brought in on each flight.[101]

- AWADS was used for the first time since Vietnam during allied airdrop operations in Eastern Bosnia. There, airdrop accuracy was increased even further by a Global Positioning System (GPS) receiver, which provided an aircraft's exact longitude and latitude and information on winds at drop altitude. However, whereas these systems are very effective for traditional, low-altitude (e.g., 1500 ft) parachute drops, in Bosnia the concern over shoulder-fired infrared missiles had forced the C-130s to drop from altitudes between 10,000 and 24,000 ft. At these altitudes, the potential for substantial wind drift was great. Since winds at lower altitudes can vary greatly, it was enormously difficult to predict the exact landing spot. Thus, for accurate parachute delivery from medium altitudes, the parachute itself needs a guidance system, not just the aircraft.[102]

Command and Control:

- Periodic breakdowns in radio communications prevented German meteorologists from obtaining eyewitness information

---

[99]Overall, three U.S. Blackhawk helicopters were shot down by RPG fire in Somalia. A 101st Division Blackhawk attached to the Quick Reaction Force was brought down on August 25, 1993, killing three crewmembers. See Jones, 1996, p. 13.

[100]Jeffrey M. Lenorovitz, "U.N.-Sponsored Airlift to Sarajevo Succeeds in Face of Frequent Attacks," *Aviation Week & Space Technology,* July 27, 1992, pp. 60–61.

[101]Although C-5s could have landed at Sarajevo, airlift organizers were concerned that the planes' huge cargo load and servicing requirements might overwhelm ground crews. Steve Vogel, "Provide Promise Takes Supplies to Sarajevo," *Air Force Times,* Vol. 52, July 20, 1992, p. 16.

[102]Tony Capaccio, "Bosnia Airdrop," *Air Force Magazine,* July 1993, pp. 52–55; and Julie Bird, "Airdrops Set for Remote Regions of Bosnia," *Air Force Times,* March 8, 1993, p. 8.

on weather conditions in the vicinity of Stalingrad, further complicating resupply operations there.[103]

- The failure of communications was, in large part, responsible for the lack of reinforcement and resupply (as well as of close air support) provided the British 1st Division at Arnhem. The range of battalion, brigade, and divisional radios was insufficient, and some signallers were inadequately trained. Moreover, there were misunderstandings regarding frequencies, call signs, codes, etc.[104]

- At Bastogne, the capability of airborne pathfinders, equipped with fluorescent panels, beacons, radios, and radar, in guiding the delivery of the initial airdrops and in warning ground crews of the approach of Allied planes helped to ensure the success of the resupply effort.[105]

- A command rivalry between the U.S. European Command (USEUCOM) and the U.S. Strike Command (USSTRICOM) left the latter with only 48 hours of planning time prior to the airborne assault on Stanleyville in 1964. On the plus side, the U.S. and Belgian militaries mostly operated very well together, even though the two had never before conducted combined airborne operations.[106]

- By contrast, the ground convoy en route to Stanleyville was unable to report a delay in its planned link-up with paratroopers in that city, which probably reduced the effectiveness of the DRAGON ROUGE operation.[107]

- In Bosnia, the United Nations had the day-to-day responsibility for nominating airdrop targets, which meant that the JFACC often had insufficient lead time to task national assets for current target information. The need to integrate the French and

---

[103]Craig, 1973, p. 242.

[104]Christopher Hibbert, *The Battle of Arnhem*, London: Batsford, 1962, pp. 204–205.

[105]Doubler, 1994, p. 220.

[106]Odom, 1988, pp. 156–157.

[107]Other C2 difficulties arose from a communications system that was overloaded with classified information and from the five different languages being spoken by friendly forces. Odom, 1988, pp. 158–159.

German allies into the Bosnian operation, as well as the lack of standardized equipment for communications, navigation, and targeting, further complicated the U.S.–led resupply effort.[108]

Political Factors:

- For a long time, the Soviet government resisted dropping arms to the Polish Home Army during the WW II Warsaw rebellion, and it refused to allow British and U.S. planes to land on airfields under its control, dooming whatever slim chance the Allied airdrop operation had to succeed.[109]

- U.S. and Belgian politicians in 1964 were sensitive to charges, by the Eastern bloc and other African states, of neo-colonialism with regard to the Congo, which delayed the initial airborne rescue operation in Stanleyville. After Paulis, that sensitivity was an important factor in preventing any further attempts to rescue European and American citizens caught up in the Congolese civil war.[110]

- At the end of the Vietnam War, the U.S. ambassador's decision to delay the final evacuation of American citizens from Saigon left little time to notify and assemble those wishing to depart. Although planners provided for only a very limited number of helicopter sorties from the U.S. Embassy building, several thousand would-be evacuees showed up at that location, swamping the available transport.[111]

- After six weeks of failure and a couple of bungled attempts, the commander of the Ranger Task Force in Mogadishu was under

---

[108]James J. Brooks, *Operation Provide Promise: The JFACC's Role in Humanitarian Assistance in a Non-Permissive Environment—A Case Study*, Newport, R.I.: Naval War College, June 14, 1996, pp. 4–5.

[109]Wilfred P. Deac, "City Streets Contested," *World War II*, Vol. 9, September, 1, 1994, pp. 43–44.

[110]Wagoner, 1980, pp. 189–191.

[111]Prior to the final evacuation, aircraft and ships often departed South Vietnam only partially filled with passengers. One reason was Ambassador Graham Martin's preference for avoiding a dramatic expansion of the evacuation effort that might demoralize the South Vietnamese military and cause it to become uncooperative with U.S. authorities. U.S. Air Force, Headquarters PACAF, 1978, pp. 127–128. See also Todd, 1990, pp. 366–367.

severe political pressure to capture warlord Aideed. This pressure probably contributed to the risky daylight air assault against the Somali leader's stronghold near the Bakara Market.[112]

Intelligence:

- At Arnhem, confusion over the location of German armored units and a mistaken belief that gliders could not land on polder (drained marshland) contributed to the British 1st Airborne Division's unfortunate selection of a staging site that was not only far from the intended target but turned out to be a rest area for an entire Panzer division.[113]

- By contrast, inaccurate intelligence led the 82nd Airborne Division to plan for substantial armored resistance at Nijmegan, thereby assisting General Gavin's troops in dealing with less formidable German opposition.[114]

- Because of the U.S. Embassy's initial overstatement of the threat to U.S. citizens and property in Monrovia, U.S. Marines received sufficient training and preparation time to carry off a remarkably smooth NEO when one became necessary during the summer of 1990.[115]

- While not fatal to the operation, initial uncertainty about the location of the U.S. Embassy in Somalia hampered rescuers involved in the 1991 Mogadishu NEO. The embassy had moved from the center of Mogadishu to the suburbs 18 months before, but Marine amphibious forces in control of rescue helicopters had only a 1969 map of the city. Fortunately, the lead CH-53E pi-

---

[112]Bowden, 1999, pp. 23–27.

[113]Hibbert, 1962, pp. 202–203; and Alexander McKee, *The Race for the Rhine Bridges, 1940, 1944, 1945*, New York: Stein and Day, 1971, p. 144. In a retrospective interview, General John W. Vogt recalled that he lost half of his fighter-bomber squadron trying to protect paratroopers pinned down in the Arnhem drop zone. See Richard H. Kohn and Joseph P. Harahan, eds., *Air Interdiction in World War II, Korea, and Vietnam*, Washington, D.C.: U.S. Air Force, Office of Air Force History, 1986, pp. 36–38.

[114]McKee, 1971, pp. 144–145; and Tugwell, 1975, p. 22.

[115]Sachtleben, 1991, pp. 78–79.

lots were able to spot the embassy compound from the air after flying around town for 15 minutes.[116]

Tactics and Training:

- The Luftwaffe refused to allow army quartermasters to supervise the loading of transports. As a result, famished German soldiers at Stalingrad opened cases of worthless goods.[117]

- During Operation Market-Garden, British and U.S. paratroopers employed different airborne landing tactics, with markedly different results. At Arnhem, airborne landings were spread out over three days, and 1st Division paratroopers were dropped 6 to 8 miles from their intended target of Arnhem Bridge.[118] By contrast, most of the 82nd Airborne infantry and artillery arrived in Nijmegan together, and the drop zone was sited on the main target.[119]

- The two Congo operations also featured a crucial difference in tactics. During the initial operation at Stanleyville, Belgian paratroopers waited for armored Jeeps to arrive at the airfield before moving into town, thus providing the rebels' time to begin killing their hostages. At Paulis, one company of paratroopers immediately moved out in search of hostages, probably saving a number of lives.[120]

- At An Loc, the USAF's use of GRADS in an attempt to resupply the town from high altitudes was initially stymied by ill-trained Vietnamese parachute packers. Until the U.S. Army sent in packing specialists from Okinawa, parachutes regularly malfunctioned, causing bundles to drift outside the drop zone and into enemy hands.[121]

---

[116]Siegel, 1992, pp. 23–24.

[117]These goods included tons of marjoram and pepper, and millions of contraceptives. Craig, 1973, p. 242.

[118]Hibbert, 1962, pp. 201–202; and Tugwell, 1975, pp. 23–28.

[119]Gavin, 1978, pp. 155–161.

[120]Wagoner, 1980, pp. 187–188.

[121]U.S. Air Force, Headquarters PACAF, 1973, pp. 28–29; and U.S. Air Force, Headquarters PACAF, 1972, pp. 25–26.

- To a large extent, a high state of readiness and advanced training in NEO procedures accounts for successful results at Mogadishu and Monrovia in the early 1990s. In Liberia, a Marine advance team, in coordination with U.S. Embassy personnel, physically sited and evaluated all potential helicopter landing zones and evacuation assembly areas.[122]

- The proximity of Desert Storm to the Mogadishu NEO meant that the requisite ships and aircraft were already in the vicinity of Somalia. In addition, the well-trained deck crews on the Marine amphibious ship *Guam* moved transport helicopters quickly and efficiently. Finally, the U.S. Embassy staff in Somalia was fully prepared to cooperate with Marine and SEAL personnel in the evacuation.[123]

- In their months-long hunt for warlord Aideed, the members of Task Force Ranger had conducted the same basic operation in Mogadishu six times previously,[124] which contributed to the loss of surprise in the October 3 air assault. In addition, "the Nightstalkers" of the 160th Special Operations Aviation Regiment (SOAR) specialized in high-speed, low-flying nighttime missions, not in hovering over congested urban areas in mid-afternoon.[125]

- Task Force Ranger's most serious tactical planning failure was the inadequate size of its rescue force. The crew of the one CSAR bird placed on standby in case of a helicopter crash was instrumental in rescuing personnel in the first downed Blackhawk. Nonetheless, there was no immediate reaction force to assist the members of Michael Durant's crew when his Blackhawk was brought down 20 minutes later.[126]

---

[122]Sachtleben, 1991, p. 81.

[123]Siegel, 1992, p. 42.

[124]The only significant tactical variation was that sometimes the forces would arrive by helicopter and leave in vehicles, and sometimes they would arrive in vehicles and leave on helicopters. Bowden, 1999, p. 23.

[125]Bowden, 1999, pp. 21, 79, 89.

[126]Bowden, 1999, p. 338.

Opposition Countermeasures:

- During the German airlift at Stalingrad, the Soviets reinforced their anti-aircraft batteries and sent fighters to harass German supply planes.[127]

- During the Warsaw Uprising, German AAA made precision airdrops by the Allies impracticable. Further complicating precision, the insurgents occupied only small and widely spread enclaves by the time the resupply operation got seriously under way.[128]

- Although enemy air defenses managed to knock down a number of airlifters, the first phase of the air resupply effort at Bastogne was assisted by the Germans, who directed very little fire on the drop zone itself.[129] During the second airlift operation, heavy flak and smoke, presumably employed by the Germans, obscured the drop zone, preventing 9 planes in the 50-aircraft serial from effectively releasing their gliders over the target area.[130]

- The poor marksmanship of airfield defenders facilitated U.S. air transport operations in the Congo in 1964.[131]

- At An Loc, the low-altitude Container Delivery System (CDS), initially employed to provide supplies to the besieged population, relied on surprise and limited adversary air defenses. However, the Communists quickly positioned AAA, and later SA-7 missiles, on all possible air approaches to the town. As a

---

[127]Latimer (1985, p. 48) attributes most of the success in interdicting the German air resupply effort to the constant operation of Soviet pursuit planes. See also Craig, 1973, p. 237.

[128]Garlinski, 1985, pp. 292–293.

[129]Marshall, 1988, p. 137.

[130]Still, it was estimated shortly after the fact that 82 percent of the supplies dispatched by the 50th Troop Carrier Wing on December 23 and 26, 1944, were delivered to the Bastogne defenders. Headquarters, 50th Troop Carrier Wing, Office of the A-2, "Analysis of Bastogne Resupply by Units of This Command," Maxwell AFB, Ala.: U.S. Air Force Historical Research Agency, January 18, 1945, pp. 2–3.

[131]Wagoner (1980, pp. 187, 197), an analyst of the Congo rescues, refuses even to speculate about what might have happened had the rebel Congolese soldiers effectively employed the automatic weapons emplaced around Stanleyville's airport during the airlift phase of DRAGON ROUGE.

result, several allied aircraft were lost, and transport planes were forced to fly above 10,000 ft in order to survive.[132]

- In Mogadishu, Aideed's forces possessed seemingly unlimited supplies of ammunition and hundreds of RPGs in the early 1990s, rendering Blackhawk operations extremely hazardous. This danger was exacerbated by the adversary's perceptions of the political advantages of shooting down U.S. helicopters and dragging dead American soldiers through the streets in full view of the international media.[133]

- During the Bosnian War, the Serbs possessed many large-caliber weapons that could have been employed to disrupt the international air resupply effort in Sarajevo. This potential was particularly serious because the Serbs controlled the mountainous terrain surrounding the Bosnian capital. Although small-arms fire remained a problem, the threat of retaliation by the NATO allies seems to have constrained the Serbs from targeting airlifters with AAA and SAMs.[134]

Atmospheric and Light Conditions:

- Low clouds, fog, and blizzards appeared in the vicinity of Stalingrad in 1942, forcing German transport aircraft to detour to bases hundreds of miles away and leaving the increasingly isolated 6th Army without access to the planes' cargoes for several days at a time. This problem was exacerbated by winter ice— which tore up aircraft engines—and cold—which made it difficult for mechanics to perform necessary aircraft maintenance.[135]

- Bad weather or a full moon during the Warsaw Uprising caused a majority of nights (37 out of 60) to be lost to Allied airdrop operations.[136]

---

[132]U.S. Air Force, Headquarters PACAF, 1973, pp. 29–30.

[133]Stevenson, 1995, p. 102.

[134]During the airlift, 93 aircraft were fired on. Only one, an Italian transport, was shot down and its crew of four killed as it was headed into Sarajevo. Arana-Barradas, 1996, p. 45; and Lenorovitz, 1992, p. 60.

[135]Craig, 1973, p. 242.

[136]Garlinski, 1985, p. 295.

- A break in the weather during the second week of the siege of Bastogne permitted effective air resupply to commence.[137]

- At An Loc, nighttime operations involved less risk to aircrews from hostile anti-aircraft fire than operations occurring during the day. However, they were generally less accurate, and ground parties and FACs had trouble observing where the packages were falling.[138]

- By the time of the Bosnian War, night and adverse weather permitted allied transport aircraft to hide from optically or IR-guided Serbian air defense weapons without degrading airdrop accuracy. However, deep snow made it difficult for besieged Muslims to recover fallen parcels.[139]

Geography and Terrain:

- Warsaw was at the maximum range of Anglo-American bombers attempting to resupply Polish insurgents during WWII. Without Soviet landing rights, Allied aircraft were forced to return to bases in Italy during daylight, over German-occupied Hungary and Yugoslavia, where the heavy planes were easy pickings for Nazi fighters.[140]

- The availability of a large, clear, gently sloping field directly west of town was a positive factor in successful U.S. resupply operations at Bastogne.[141]

- The small size of the drop zone at An Loc meant that many airdrops that just missed the target ended up in the hands of the enemy.[142]

---

[137]Charles B. MacDonald, *A Time for Trumpets: The Untold Story of the Battle of the Bulge,* New York: William Morrow, 1985, p. 521.

[138]U.S Air Force, Headquarters PACAF, 1973, p. 30.

[139]Steve Vogel, "Bosnia Airdrop Crews Glad for Dark, Clouds," *Air Force Times,* March 15, 1993, p. 4.

[140]Overall, the Allies lost about 13 percent of the aircraft that flew in support of the Warsaw Uprising. Garlinski, 1985, pp. 285–287.

[141]Marshall, 1988, p. 135.

[142]U.S. Air Force, Headquarters PACAF, 1973, p. 31.

- During the Saigon NEO, the U.S. Embassy rooftop, which was used throughout the embassy evacuation as a landing zone for CH-46 transport helicopters, could not support the weight of the larger CH-53s. Consequently, CH-53 operations took place in the embassy parking lot. This location and the larger-than-expected number of evacuees caused the evacuation process to be extended far beyond the planned completion time.[143]

- The location of the U.S. raid on "Bloody Sunday," in the heart of Habr Gidr clan's territory in central Mogadishu, led to the unraveling of the operation after Task Force Ranger became pinned down. With hundreds of thousands of clan members living in the vicinity, it was one of the few places where Aideed's forces could quickly mount a serious fight. In addition, the urbanized terrain of densely packed buildings and narrow streets offered few landing zones large enough for helicopters to extract ground troops.[144]

- In Bosnia, the mountainous terrain limited the number of suitable drop zones while providing good cover for forces opposed to the resupply operation.[145]

INTERDICTION AND SIEGE SUPPORT

Although not often considered as instruments of urban operations, air interdiction and aerial siege support have affected the outcome of city battles from Leningrad to Khafji. When successful, interdiction[146] has helped to isolate the urban battlefield, metering or disrupting the flow of opposition reinforcements and supplies and providing friendly forces with the long-term advantage in the close-in battle. Historically, effective urban interdiction operations have required air superiority, an abundance of available bombers, good weather, moderately open terrain, and a mechanized opposition force with long and constricted lines of communication (LOC). Major

---

[143]U.S. Air Force, Headquarters PACAF, 1978, p. 153.

[144]Bowden, 1999, pp. 20–21.

[145]Brooks, 1996, p. 12.

[146]*Interdiction* is attacks on enemy lines of communication to slow or stop the movement of vehicles, personnel, and supplies.

factors inhibiting interdiction in the past have included the ability of some opponents to off-load supplies onto ever smaller conveyances and to perform logistics operations under the cover of darkness. However, the development of precision air-ground weapons, advanced C4ISR systems, and nighttime attack capabilities has added significantly to interdiction's effectiveness in recent years—to the point where coalition ground forces during the Battle for Khafji were able to quickly turn back a two-brigade Iraqi attack.

By contrast, aerospace power has had only moderate success as an instrument of siege warfare. Barring a sudden and massive attack on a city, urban residents generally appear to become accustomed to the terror and destruction of aerial bombing, sometimes to the extent that their suffering becomes a source of pride, fueling their desire to resist. This appears to have been the case during the Germans' 3-year siege of Leningrad during World War II. The experience of the Israelis with aerial bombardment during the 1982 siege of Beirut was somewhat more positive, possibly because they used aerospace power in a more discriminating fashion: as a means for dividing PLO fighters from the local Lebanese population. Still, Israel lacked the forces and the will to destroy the PLO through conventional bombing alone.

Results

Attempts by both sides in the European theater during WWII to interdict the supply lines of forces moving toward urban combat zones proved more effective on the Western Front than on the Eastern Front. Although the Luftwaffe strafed and bombed Leningrad's "ice bridge" (a frozen lake that supplies were driven across in winter), particularly where large fissures in the ice caused a pileup of supply vehicles, it failed to halt traffic for long or, more important, to destroy the loading and unloading facilities on Lake Ladoga's shores.[147] At Stalingrad, German dive-bombers disabled or sunk many Russian ferries used to transport troops and supplies across the Volga River. In addition, German artillery and aircraft struck Russian footbridges,

---

[147]Gouré, 1962, p. 152.

making daytime river crossings nearly impossible. Nonetheless, nighttime resupply and reinforcement continued.[148]

The impact of interdiction on the Anglo-American campaigns in Normandy and the Ardennes in 1944–1945 was quite different. As the Allies established themselves in Normandy, air interdiction slowed the advance of German armored reinforcements to a crawl, primarily by destroying French rail centers and bridges.[149] Without interdiction's help, it is questionable whether the Allies could have overcome stubborn German resistance at places such as Brest, Cherbourg, and Caen. Likewise, aerospace power disrupted German LOCs in the vicinity of the Ardennes, relieving pressure on Allied ground units, such as the 101st Division at Bastogne. Over 1,000 heavy bombers targeted railroad bridges and marshaling yards behind the German lines, almost wiping out the region's rail system. In addition, Allied fighter-bombers knocked out hundreds of enemy armored vehicles on their way to the front.[150]

Subsequent efforts by the U.S. Air Force to support urban operations by striking enemy forces located on city approaches have proven moderately successful. In An Loc, despite considerable efforts to spot and destroy enemy artillery being moved into position, a large force arrived on the town's periphery undetected and was employed with great effectiveness by the Communists during the siege. By contrast, U.S. air attacks against enemy forces interdicting Highway 13 and blocking a South Vietnamese relief column from reaching An Loc eventually paid off. In particular, a B-52 ARCLIGHT strike caught elements of the North Vietnamese Army's 7th Division in the open and obliterated them, permitting the ARVN 46th regiment to enter the town.[151] During the Persian Gulf War, coalition aerospace forces successfully thwarted the Iraqis' attempt to move reinforcements at

---

[148]Latimer, 1985, pp. 52–53; and Craig, 1973, p. 161.

[149]Even when enemy troops reached the front, they arrived too tired and demoralized from the bombing to immediately take up their positions. See Eduard Mark, *Aerial Interdiction: Air Power and the Land Battle in Three American Wars*, Washington, D.C.: Center for Air Force History, 1994, pp. 245–250.

[150]Hughes, 1995, p. 283.

[151]Still, even after the siege of An Loc was broken, small pockets of Communist forces continued to sporadically interdict the highway, forcing aerial resupply to continue. U.S Air Force, Headquarters PACAF, 1973, pp. 63–66.

night toward the Saudi town of Khafji, thus avoiding a potentially large and bloody battle over the city. According to a soldier from the 5th Iraqi Mechanized Division—one of two second-echelon divisions employed—his brigade underwent more damage in 30 minutes from allied aerospace power than it had in eight years of the Iran-Iraq War. The Iraqi battalion that did manage to get into the city either withdrew or surrendered to Arab coalition forces two days after entering Khafji.[152]

In contrast to interdiction, modern militaries have not often used aerospace power to support urban sieges. When they have, the results have been mixed at best. Having been ordered by Hitler not to directly assault Leningrad, the Wehrmacht sought to win the city through a siege that denied food and supplies to the residents and by air and artillery bombardment. Although the raids and shellings directed at the Russian city were not intensive by World War II standards, they were spaced out to interfere as much as possible with the activities of the Russian inhabitants. Moreover, they caused substantial damage to the city's industrial installations and killed and wounded many civilians, especially factory workers. Still, following the initial shock, Leningrad's population rather quickly managed to adjust to the dangers of German bombardment.[153] During the Lebanon War of 1982, the Israelis were somewhat more successful with their modified siege operations against PLO strongholds in Beirut, which featured air strikes, overflights, and other forms of aerial intimidation. Although the majority of the Palestinian fighting force survived the siege, Israel achieved its operational objective of expelling the PLO from Lebanon. However, the Israeli Defense Forces suffered almost a quarter of their total campaign losses in Beirut, despite their refusal to take part in extensive house-to-house fighting. Furthermore, the prolongation of the siege, combined with rising casualties, created public pressure on the Israeli cabinet to halt the conflict and pull Israeli troops out of central Lebanon.[154]

---

[152]Thomas A. Keaney and Eliot A. Cohen, *Gulf War Air Power Survey Summary Report*, Washington, D.C.: U.S. Department of the Air Force, 1993, p. 109.

[153]Gouré, 1962, pp. 100–105.

[154]Richard A. Gabriel, *Operation Peace for Galilee: The Israeli-PLO War in Lebanon*, New York: Hill and Wang, 1984, pp. 167–168.

Effectiveness Factors

The following is a list of factors that have contributed to the effectiveness (or ineffectiveness) of the preceding urban interdiction and siege-support operations. They are grouped into performance categories: weapons and equipment, command and control, intelligence, tactics and training, ground-force cooperation, opposition countermeasures, atmospheric and light conditions, and geography and terrain. In general, the results indicate that the technical ability of first-class aerospace forces to conduct urban interdiction and siege-support operations has improved considerably since the 1980s. Furthermore, incomplete intelligence information and misguided tactics have decreased operational effectiveness in recent decades, but usually not to a fatal degree. Finally, although quite significant during WWII, performance categories such as ground-force cooperation, opposition countermeasures, and atmospheric and geographic conditions appear to have mattered less in recent decades.

Weapons/Equipment:

- Once the Eighth Air Corps had been withdrawn from the battle, German forces outside Leningrad were left with few dive-bombers and only about 300 planes of all types. As a result, German aircraft could not bomb with sufficient intensity to significantly disrupt Soviet resupply operations.[155]

- By contrast, with the exception of the British Bomber Command, every Allied air force spent considerable effort in the early weeks of Operation Overlord interdicting German forces. For example, during the first half of June 1944, Eighth Air Force strategic bombers devoted almost all their sorties to tactical interdiction.[156]

- Still, Anglo-American aircraft during WWII had to expend a substantial amount of ordnance to destroy fixed interdiction targets

---

[155]Gouré, 1962, p. 99.
[156]Hughes, 1995, p. 149.

such as bridges.[157] Fighter-bombers, such as the P-51 and the A-36, were more accurate than bombers and could fly when the latter were grounded by weather; however, they lacked the bombload capacity to destroy massive targets and were vulnerable to light-caliber AAA when dive-bombing.[158]

- In Beirut, the Israeli Air Force largely accomplished its delicate mission of pressuring the PLO to leave while avoiding substantial collateral damage, through a careful targeting process involving the use of aerial photographs, highly accurate Maverick missiles, and small iron bombs.[159]

- By permitting accurate attacks from medium altitudes (13,000 to 30,000 ft), the Persian Gulf War confirmed the superiority of PGMs over dumb bombs. Although the GBU-12 constituted nearly 50 percent of all smart bombs dropped by American forces, the Maverick missile, fired primarily from A-10s, proved highly effective in interdicting Iraq's mechanized forces outside of Khafji.[160]

Command and Control:

- By the time of the Ardennes counteroffensive, IX TAC had developed a highly efficient control system for fighter planes. That system included forward director posts, radar centers, fighter

---

[157]According to U.S. Air Force General John Vogt, a 450-ft circular error probable (CEP) was considered good. See Kohn and Harahan, 1986, p. 9.

[158]F. M. Sallager, *Operation STRANGLE (Italy, Spring 1944): A Case Study of Tactical Air Interdiction,* Santa Monica, Calif.: RAND, R-0851-PR, February 1972, pp. 34–35; and Martin van Creveld with Steven L. Canby and Kenneth S. Brower, *Air Power and Maneuver,* Maxwell AFB, Ala.: Air University Press, July 1994, p. 194.

[159]Nevertheless, IAF operations did cause a significant number of civilian casualties, in part because many Lebanese residents of West Beirut chose to remain in their homes (in spite of Israeli warnings to evacuate), out of fear that their property might be stolen. See Gabriel, 1984, p. 160.

[160]See the interviews with General Charles Horner and Maj. General Thomas Olsen (CENTAF commander and deputy commander, respectively, during Operation Desert Storm) and Major Michael Edwards (Operation Desert Storm A-10 pilot) in Major Daniel R. Clevenger, study director, *"Battle of Khafji," Air Power Effectiveness in the Desert,* Vol. 1, Washington, D.C.: U.S. Air Force, Studies and Analyses Agency, July 1996, pp. 57, 69–70, 82; and Richard Hallion, *Storm over Iraq: Air Power and the Gulf War,* Washington, D.C.: Smithsonian Institution Press, 1992, p. 203.

control stations, and a combat command.[161] Furthermore, Maj Gen Pete Quesada's fighter command was the first to integrate the Microwave Early Warning (MEW) radar and the SCR-584 anti-aircraft radar to provide navigation and precise control to fighter-bombers during ground-attack missions.[162]

- Khafji provides an example of how a well-oiled command and control system can help ensure successful urban interdiction operations. Once Iraqi offensive intentions were apparent, the Coalition Air Operations Center moved quickly to redirect already-scheduled sorties toward moving enemy forces. A i r attacks were funneled into the Kuwaiti Theater of Operations (KTO) from different altitudes and directions using a grid of designated "kill boxes"[163] as a control measure.[164]

- Much of the night-interdiction effort in southern Kuwait was directed by Marine Fast FACs in F/A-18D aircraft, who identified and marked targets for other planes carrying weapons. Having penetrated over 5 miles inside Kuwait, a Marine Air/Naval Gunfire Liaison Company (ANGLICO) team also called in air strikes against hundreds of Iraqi vehicles preparing to move south.[165]

---

[161]Hughes, 1995, p. 294.

[162]The wide-band MEW was used for long-range and area control; the SCR-584, with its narrow beam, was used for close-range, precision work. Radar operators helped fighters get under and through the weather both in the target areas and at recovery bases and validated targets by correlating ground locations with tracked fighter positions. Quesada's command used the SCR-584 to blind-bomb through overcast skies and to direct aerial reconnaissance flights. However, the blind-bombing method had some problems. Small shifts in temperature had a significant effect on delicate ground radar equipment, creating the potential for large bombing errors on the battlefield. Furthermore, the process of entering bombing data into the control stations proved too lengthy. See William R. Carter, "Air Power in the Battle of the Bulge: A Theater Campaign Perspective," *Airpower*, Vol. 3, Winter 1989, pp. 26–27; and Hughes, 1995, p. 294.

[163]The KTO was divided into many zones, called kill boxes, to organize and control air attacks against forces in somewhat featureless terrain.

[164]Planners managed to push a four-ship flight of aircraft through each kill box every 7 to 8 minutes during the day and every 15 minutes at night. Grant, 1998, p. 31; and Clevenger, 1996, pp. 20–21.

[165]Rebecca Grant, 1998, pp. 31–32.

Intelligence:

- Following the withdrawal of the South Vietnamese army from the main fire-support base outside An Loc in 1972, the surrounding area was left devoid of friendly ground troops, which severely hampered allied intelligence efforts. USAF forward air controllers and remaining elements of the U.S. 1st Air Cavalry regiment were spread so thinly that they could provide little definite information about the locations of the three Communist divisions moving in the direction of the provincial capital.[166]

- Subsequently, however, South Vietnamese intelligence sources provided accurate information on Communist plans to intercept an ARVN unit coming south from An Loc to assist forces attempting to relieve the town. This information resulted in a B-52 ARCLIGHT strike that totally decimated a North Vietnamese regiment.[167]

- In Khafji, U.S. JSTARS MTI sensors detected and recorded the initial preparations for movement of Iraq's 5th Mechanized Division and 3rd Armored Division before they crossed the Kuwaiti-Saudi border. Apparently, however, coalition analysts did not at first understand the significance of the data; only later did it become clear that the Iraqi buildup portended an attack on the town of Khafji. Nonetheless, once the invasion had begun, JSTARS' ability to detect and pass along information on Iraqi reinforcements proved to be an essential element in the coalition air force's effort to isolate Iraqi units inside the town.[168]

Tactics and Training:

- In Beirut, the Israelis made a distinction between PLO-controlled areas and camps in the southwest and the northwestern part of the city, where Lebanese Sunnis predominated. For example, PLO areas were subjected to numerous flyovers, flare drops, and

---

[166]U.S. Air Force, Headquarters PACAF, 1973, p. 2.

[167]U.S. Air Force, Headquarters PACAF, 1973, p. 53.

[168]During Desert Storm, JSTARS was most useful for providing overall situational awareness. However, precise targeting with JSTARS was difficult, because of the lack of a reliable, accurate interface with coalition attack assets. Clevenger 1996, pp. 56, 65.

sonic booms intended to intimidate the families of PLO members.[169]

- With the notable exception of the massive aerial bombardment of PLO camps on August 12, 1982, the IAF did not carry out systematic terror bombing, even in areas of Beirut with a Palestinian majority. The exception to the rule resulted in an enormous public outcry, both internationally and domestically, against Israeli policies in Lebanon. The outcry prevented Israeli Defense Minister Ariel Sharon from realizing his personal vision of destroying the PLO as a fighting force.

Ground-Force Cooperation:

- The Finns' refusal to close the Lake Ladoga corridor to Leningrad early on, and the subsequent failure of the German Tikhvin offensive, probably doomed any chance Germany had to halt the flow of supplies to Russia's second-most important city.[170]

- In Beirut, the relative inexperience of the Israeli army in urban warfare, and the refusal of the Lebanese Christian faction to take on the bloody job of house-to-house fighting, led Israeli commanders to pursue a modified siege strategy with respect to the PLO. In such a situation, aerospace power came to play a central role in driving Palestinian forces out of Lebanon.[171]

- For Khafji, friendly ground forces in the area were limited to primarily border reconnaissance teams and the U.S. Marine Task Force Shepherd, a two-battalion screening force[172] for the 1st Marine Division down at Kibrit.[173] Coalition air forces did provide close support to these and other friendly ground forces, and

---

[169]Gabriel, 1984, pp. 136–139, 157–159. See also R. D. McLaurin and Paul A. Jureidini, *The Battle of Beirut, 1982*, Aberdeen Proving Ground, Md.: U.S. Army Human Engineering Laboratory, January 1986, p. 48.

[170]Paul Carell, *Hitler Moves East, 1941–1943*, Boston: Little, Brown, 1963, p. 267; and John Erickson, *The Road to Stalingrad: Stalin's War with Germany*, Vol. 1, New York: Harper & Row, 1975, pp. 270, 277–278.

[171]McLaurin and Jureidini, 1986, p. 30; and Gabriel, 1984, pp. 130–132.

[172]A *screening force* is put in front of or to the side of a main force to provide early warning of and some defense against a major enemy attack.

[173]Atkinson, 1993, p. 198.

the ground combat units and reconnaissance teams also assisted the interdiction effort by directing fire on enemy forces not yet in contact. That said, the interdiction effort was largely an independent air operation with JSTARS and airborne FACs (both Marine and USAF) directing most strike aircraft in on the Iraqi 5th Mechanized and 3rd Armored Divisions as they moved south toward Khafji.

Opposition Countermeasures:

- To counter direct enemy bombardment during the siege of Leningrad, the Soviets relied on significant air defenses, including over 100 fighters, numerous anti-aircraft guns, barrage balloons, and searchlights.[174]

- The success of the Soviet "ice bridge" operation, in the face of German bombardment and severe weather, was ensured only after the truck convoy system was abandoned. That system hampered drivers willing to make several trips in a row across Ladoga. Egged on by local Communist Party, Komsomol, and NKVD officials,[175] individual truck drivers were able to make as many as four round trips a day during shifts lasting from 16 to 18 hours.[176]

- At Stalingrad, German attempts at interdicting the Soviet LOC across the Volga and the Don rivers were hindered by several Soviet countermeasures, including protecting the railway lines with fighters and AAA, unloading supplies from trains onto trucks as far as 150 miles from the front, employing large numbers of troops to hand-carry supplies, and constructing pontoon bridges just beneath the river's surface to hide them from accurate artillery fire and dive bombers.[177]

---

[174]Gouré, 1962, p. 99.

[175]The Komsomol was the youth branch of the Soviet Communist Party (literally, the Young Communist League). The NKVD was a Stalin-era Soviet intelligence and internal security organ, a predecessor to the KGB.

[176]Moving supply bases forward and extending the railroad to the lake shore also speeded up the resupply operation and countered the effects of German interdiction. Gouré, 1962, pp. 206–209.

[177]Erickson, 1975, p. 411; Craig, 1973, p. 161; and Latimer, 1985, pp. 52–53.

- In 1982, the Palestinians had no air forces and insignificant numbers of AAA and SAMs with which to confront the Israeli Air Force in Beirut. They were able to compensate somewhat for these shortcomings through the use of clever tactics. For example, after deliberately placing its positions in civilian areas, the PLO provoked outrage among the international public and in Israel when Israeli Air Force bombs killed civilians or destroyed homes.

- In addition, the Palestinians in Beirut protected themselves from air attack by keeping their units small and highly mobile and by constantly changing their locations. Moreover, having had plenty of time to prepare for Israeli bombardment, they developed an extensive network of underground tunnels and trenches.[178]

Atmospheric and Light Conditions:

- Bad weather turned out to be a more effective interdiction asset against the Soviet resupply operation across Lake Ladoga than German air or artillery. For example, snowstorms and blizzards occurred on 22 days during the month of February 1942, requiring nearly constant snow-removal operations to keep the ice road open.[179] Nevertheless, the Soviets could not have survived the German siege if Ladoga had remained unfrozen.

- The weather worked against the German interdiction effort at Stalingrad. When the Soviets finally broke through the Don River barrier, both Soviet and German air forces were grounded by the weather.[180]

- During the initial period of the Ardennes offensive, low cloud ceilings and snow prevented Allied fighter-bombers from making any substantial strikes against German columns. But as soon as the clouds cleared, U.S. and British planes took to the sky in large

---

[178]Gabriel, 1984, pp. 132–133, McLaurin and Jureidini, 1986, p. 33.

[179]In all, 1,004 Russian trucks were smashed or lost while navigating the road, the majority due to the weather, and most trucks required repairs after each trip. Gouré, 1962, p. 152.

[180]Craig, 1973, p. 187.

numbers, just when German supply lines were stretched to the limit.[181]

- During the Khafji operation, aerospace power operated almost as effectively at night as during the day. The Low Altitude Navigation and Targeting Infrared for Night (LANTIRN)-equipped F-15E scored first-pass kills against individual Iraqi vehicles at night and in bad weather.[182] However, the venerable AC-130 gunship and the A-10 Warthog were the most lethal nighttime performers.[183]

Geography and Terrain:

- In the battles of Leningrad and Stalingrad, the Soviets possessed certain geographical advantages that frustrated German interdiction efforts. Owing to the failure of the Finns to press their initial advantage against the Red Army and link up with German forces in the south, the Soviets retained a 50-mile-wide corridor between the city of Leningrad and the far shore of Lake Ladoga. As suggested earlier, this became a significant advantage in resupplying the city once the lake froze and the "ice bridge" was constructed.

- The location of Stalingrad on the west bank of the Volga River permitted the east bank to be utilized as a fairly secure supply base and location for indirect-fire artillery. Because that artillery required vast quantities of ammunition, the Soviets may not have been able to meet their overall logistics needs in the initial period of the battle if the artillery had been forced to move across the river.[184]

- The hilly terrain of the Ardennes, traversed by narrow rural roads with few exits, benefited Allied fighter-bombers attempting to interdict German armored reinforcements. Allied pilots were able to block entire columns with solitary strikes aimed at lead

---

[181]Hughes, 1995, pp. 280–283.

[182]Hallion, 1992, pp. 314–315.

[183]Still, as the A-10 loss rate began to climb, Central Air Force (CENTAF) commander Horner greatly scaled back Warthog operations. Clevenger, 1996, p. 27.

[184]Latimer, 1985, p. 52.

vehicles, taking out the remaining enemy vehicles at their leisure.[185]

CONCLUSION

What does the historical record have to say about the overall performance of U.S. aerospace forces in past urban operations? To begin with, it must be acknowledged that all four military services have accumulated substantial experience, from World War II to Kosovo, in providing air support to joint urban operations during periods of war and relative peace. That support has included providing close air support to embattled ground troops in such diverse places as Cherbourg, Hue, and Mogadishu, as well as to friendly civilians during noncombatant evacuation operations in Saigon and Tirana. It has included providing logistics support to friendly troops and civilians in airdrop and airlift operations, as conducted during the battles of Bastogne, An Loc, and Sarajevo, and in transport operations, as occurred at Arnhem, Stanleyville, and Monrovia. Moreover, it has included air interdiction operations during the Normandy offensive and the Battle of Khafji, as well as C4ISR and psychological warfare activities in Grenada, Panama, and Somalia.

Despite this extensive record, the effectiveness of American aerospace power in urban operations has varied so much throughout the years that no general trend is discernible. With regard to close air support, Cherbourg, An Loc, and Panama can be counted as successes, and Aachen, Hue, and Grenada as failures. Whereas CAS difficulties during World War II often stemmed from the inability of existing air weaponry to destroy fortified defenses, in recent times, they have had more to do with a heightened concern over friendly military and noncombatant casualties.

As to logistics support, U.S. air forces have been successful in resupplying besieged cities, such as An Loc and Sarajevo, but have had serious problems with troop transport in cities such as Mogadishu, where the opposition possessed numerous, albeit rather unsophisticated, means of air defense. For its part, the Air Force has demonstrated considerable success in observing and interdicting the

---

[185]Hughes, 1995, p. 284.

movement of enemy forces and supplies bound for such urban battlefields as Cherbourg, Bastogne, and Khafji.

Because of the variety of examples, no simple formula for aerospace force success can be derived from past urban operations. Nevertheless, a few general historical observations can be made:

- Urban close air support has usually been easier to conduct when friendly ground forces were on the defensive (e.g., Bastogne and An Loc) rather than on the offensive (e.g., Cassino and Hue).

- Urban airdrops have at times been very precise and useful but, unless the target population was highly concentrated as at An Loc, usually have not replaced other means of resupply.

- Helicopter transport within contested urban areas has become quite hazardous.

- At least in conventional conflicts, interdiction of the approaches to a city occupied by hostile forces has often been the most effective means of aerial fire support.

- Employing aerospace forces to support siege operations, such as the Israelis did in Beirut, has become militarily feasible, but would probably be politically unwise for a democratic country like the United States.

Most of the same factors that have contributed to effective air operations in other environments have been successfully applied in urban settings as well, including the following:

- Careful mission planning to ensure that air assets are used appropriately

- The ability to suppress or circumvent opposition air defenses

- Close coordination between friendly aerospace and ground forces

- Precision weapons and tactics

- Most important, identifiable and targetable adversary forces (i.e., not too dispersed, hidden, fortified, or intermixed with civilians and/or friendly troops).

BIBLIOGRAPHY

Adkin, Mark, *Urgent Fury: The Battle for Grenada*, New York: Lexington Books, 1989.

"Air Force Operations in Support of Attack on Cherbourg," Maxwell AFB, Ala.: U.S. Air Force Historical Research Agency, June 22–30, 1944.

Alexander, John B., *Future War: Non-Lethal Weapons in Twenty-First Century Warfare*, New York: St. Martin's Press, 1999.

Altmann, J., "Cooperative Monitoring of Limits on Tanks and Heavy Trucks Using Acoustic and Seismic Signals—Experiments and Analysis," *Proceedings of the 5th Battlefields Acoustics Symposium*, Ft. Meade, Md., September 23–25, 1997, pp. 135–174.

Americas Watch, *The Laws of War and the Conduct of the Panama Invasion*, May 1990.

Amnesty International, *Israel/Lebanon: Unlawful Killings During Operation Grapes of Wrath*, London, England, July 1996.

Anderson, Jon R., "Rescue 911: On the Ground with Marines in Albania," *Navy Times*, March 31, 1997, p. 13.

Anthes, J., et al., "Non-Scanned LADAR Imaging and Applications," *Applied Laser Radar Technology, Proceedings of the SPIE*, Vol. 1936, 1993.

Arana-Barradas, Louis A., "A 'Promise' of Peace: Sarajevo Humanitarian Airlift Ends, New Hope Begins," *Airman,* March 1996.

Ashworth, G. J., *The City and War,* London: Routledge, 1991.

Atkinson, Rick, *Crusade: The Untold Story of the Persian Gulf War,* Boston: Houghton Mifflin, 1993.

Baum, C., et al., "The Singularity Expansion Method and Its Application to Target Identification," *Proceedings of the IEEE,* Vol. 79, No. 10, October 1991, pp. 1481–1491.

Beaver, Paul, "Army Aviation in Chechnya," *Jane's Defense Weekly,* June 10, 1995.

Bingham, Price, *The Battle of Al Khafji and the Future of Surveillance Precision Strike,* Arlington, Va.: Aerospace Educational Foundation, 1997.

Bird, Chris, "Kosovo Crisis: Yugoslav Media Fear Crackdown Amid War Fever," *Guardian,* October 8, 1998, p. 15.

Bird, Julie, "Airdrops Set for Remote Regions of Bosnia," *Air Force Times,* March 8, 1993, p. 8.

Birkler, John, C. Richard Neu, and Glenn Kent, *Gaining New Military Capability: An Experiment in Concept Development,* Santa Monica, Calif.: RAND, MR-912-OSD, 1998.

Blechman, Barry M., and Tamara Cofman Wittes, "Defining Moment: The Threat and Use of Force in American Foreign Policy," *Political Science Quarterly,* Vol. 114, No. 1, 1999, pp. 1–30.

Bothe, Michael, Karl Josef Partsch, and Waldemar A. Solf, *New Rules for Victims of Armed Conflicts,* The Hague: Martinus Nijhoff Publishers, 1982.

Boudreaux, Richard, "Chechens Drop Russia Talks After Leader's Death," *Los Angeles Times,* April 25, 1996.

Bovais, C., "Integration and Flight Demonstration of a Biological Warfare Agent Detection System on an Unmanned Aerial Vehicle,"

AUVSI '98, Association for Unmanned Vehicle Systems International, 1998.

Bowden, Mark, *Black Hawk Down: A Story of Modern War*, New York: Atlantic Monthly Press, 1999.

Brendley, Keith W., and Randall Steeb, *Military Applications of Microelectromechanical Systems*, Santa Monica, Calif.: RAND, MR-175-OSD/AF/A, 1993.

Brooks, James J., *Operation Provide Promise: The JFACC's Role in Humanitarian Assistance in a Non-Permissive Environment—A Case Study*, Newport, R.I.: Naval War College, June 14, 1996.

Bunker, Robert J., ed., *Nonlethal Weapons: Terms and References*, Colorado Springs, Colo.: U.S. Air Force Academy, INSS Occasional Paper 15, 1997.

Butcher, Tim, and Patrick Bishop, "NATO Admits Air Campaign Failed," *London Daily Telegraph*, July 22, 1999, p. 1.

Byman, Daniel, and Matthew Waxman, "Defeating US Coercion," *Survival*, Vol. 41, No. 2, Summer 1999.

Byman, Daniel, Matthew Waxman, and Eric Larson, *Air Power As a Coercive Instrument*, Santa Monica, Calif.: RAND, MR-1061-AF, 1999.

Callard, James R., "Aerospace Power Essential in Urban Warfare," *Aviation Week & Space Technology*, September 6, 1999.

Cantella, M., "Micro Air Vehicle Sensor," *Proceedings of the IRIS Specialty Group on Passive Sensors*, Vol. 1, 1999.

Capaccio, Tony, "Bosnia Airdrop," *Air Force Magazine*, July 1993, pp. 52–55.

———, "U.S. Snipers Enforced Peace Through Gun Barrels," *Defense Week*, January 31, 1994, p. 1.

Carell, Paul, *Hitler Moves East, 1941–1943*, Boston: Little, Brown, 1963.

Carter, William R., "Air Power in the Battle of the Bulge: A Theater Campaign Perspective," *Airpower,* Vol. 3, Winter 1989, pp. 26–27.

"The Casualties of Chechnya," *The New York Times,* August 10, 1995, p. 18.

Center for Strategic and International Studies, *Clashes of Visions: Sizing and Shaping Our Forces in a Fiscally Constrained Environment,* Proceedings of CSIS-VII Symposium held October 29, 1997, Washington, D.C., 1998.

Clevenger, Daniel R., study director, *"Battle of Khafji," Air Power Effectiveness in the Desert,* Vol. 1, Washington, D.C.: U.S. Air Force, Studies and Analyses Agency, July 1996, pp. 57–82.

Cohen, Eliot A., "The Mystique of U.S. Air Power," *Foreign Affairs,* Vol. 73, January–February 1994.

Cohen, Roger, "Fighting Rages As NATO Debates How to Protect Bosnian Enclave," *The New York Times,* November 25, 1994, p. A1.

Cole, Ronald H., *Operation Urgent Fury: Grenada,* Washington, D.C.: Office of the Chairman of the Joint Chiefs of Staff, Joint History Office, 1997.

Corsini, G., et al., "Simulated Analysis and Optimization of a Three-Antenna Airborne InSAR System for Topographic Mapping," *IEEE Transactions on Geoscience and Remote Sensing,* Vol. 37, No. 5, September 1999, pp. 2518–2529.

Craig, William, *Enemy at the Gates,* New York: Reader's Digest Press, 1973.

Crawford, Leslie, "Unrepentant Peacekeepers Will Fire on Somali Human Shields," *Financial Times,* September 11, 1993, p. 4.

Crossette, Barbara, "Civilians Will Be in Harm's Way If Baghdad Is Hit," *The New York Times,* January 28, 1998, p. A6.

Cushing, Dan E., X0-HMM-365, *Operation Silver Wake,* Oral History Interview, Washington, D.C.: Marine Corps Historical Center, Marine Corps Oral History Program, June 12, 1997.

DARPA, "Airborne Communications Node," URL: http://www.darpa.mil/ato.

David, William C., "The United States in Somalia: The Limits of Power," *Viewpoints*, Vol. 95, No. 6, June 1995, p. 9.

Davis, William R., Jr., Bernard B. Kosicki, Don M. Boroson, and Daniel F. Kostishack, "Micro Air Vehicles for Optical Surveillance," *The Lincoln Laboratory Journal*, Vol. 9, No. 2, 1996, pp. 197–214.

Deac, Wilfred P., "City Streets Contested," *World War II*, Vol. 9, September 1, 1994, pp. 43–44.

Defense Mapping Agency (now NIMA), *Geodesy for the Layman*, DMA TR 80-083, available at ftp://ftp.nima.mil/pub/gg/geo4layman/Geo4lay.pdf.

Deinekin, Pyotr S., "Where Are We Directing the Flight of Our Birds? On the Air Force's Status and Development Prospects," *Armeiskii sbornik*, August 1996, pp. 9–12.

DeSaussure, Ariane L. "The Role of the Law of Armed Conflict During the Persian Gulf War: An Overview," *Air Force Law Review*, Vol. 27, 1994, pp. 60–61.

D'Este, Carlo, *Decision in Normandy*, New York: E. P. Dutton, 1983.

Dewar, Michael, *War in the Streets: The Story of Urban Combat from Calais to Khafji*, New York: David & Charles, 1992.

Donnelly, Thomas, Margaret Roth, and Caleb Baker, *Operation Just Cause: The Storming of Panama*, New York: Lexington Books, 1991.

Dorr, L., News Release from Federal Aviation Administration Technical Center, URL: http://www.faa.gov/apa/pr/, August 13, 1999.

Doubler, Michael D., *Closing with the Enemy: How GIs Fought the War in Europe, 1944–1945*, Lawrence, Kansas: University Press of Kansas, 1994.

Ellefsen, Richard, *Current Assessment of Building Construction Types in Worldwide Example Cities*, Report prepared for Naval Surface Warfare Center, Dahlgren Laboratory, Dahlgren, Va., 1999.

Ellefsen, Richard, Bruce Coffland, and Gary Orr, *Urban Building Characteristics: Setting and Structure of Building Types in Selected World Cities*, Dahlgren, Va.: Naval Surface Warfare Center, Dahlgren Laboratory, 1977.

Erickson, John, *The Road to Stalingrad: Stalin's War with Germany*, Vol. 1, New York: Harper & Row, 1975.

Florig, Keith H., "The Future Battlefield: A Blast of Gigawatts?" *IEEE Spectrum*, March 1988, pp. 53–54.

Fontana, R., et al., "An Ultra Wideband Communications Link for Unmanned Vehicle Applications," URL: http://www.his.com/~mssi, September 1999.

Freedman, Lawrence, and Efraim Karsh, *The Gulf Conflict 1990–1991*, Princeton, N.J.: Princeton University Press, 1993.

Fulghum, David, "Air War in Chechnya Reveals Mix of Tactics," *Aviation Week & Space Technology*, February 14, 2000.

———, "Desert Storm Success Renews USAF Interest in Specialty Weapons," *Aviation Week & Space Technology*, May 13, 1991.

Futrell, Robert F., *The United States Air Force in Korea: 1950–1953*, Washington, D.C.: Office of Air Force History, 1983.

Gabriel, Richard A., *Operation Peace for Galilee: The Israeli-PLO War in Lebanon*, New York: Hill and Wang, 1984.

Gallagher, Dick, USMC OpsO, HMM-261, Oral History Interview, Washington, D.C.: Marine Corps Historical Center, Marine Corps Oral History Program, May 22, 1984.

Garlinski, Jozef, *Poland in the Second World War*, New York: Hippocrene Books, 1985.

Gavin, James M., *On to Berlin: Battles of an Airborne Commander, 1943–1946*, New York: Viking Press, 1978.

Gellman, Barton, "Allied Air War Struck More Broadly in Iraq," *Washington Post*, June 23, 1991, p. A1.

Gerwehr, Scott, and Russell W. Glenn, *The Art of Darkness: Deception and Urban Operations*, Santa Monica, Calif.: RAND, MR-1132-A, 2000.

Glenn, Russell W., *Combat in Hell: A Consideration of Constrained Urban Warfare*, Santa Monica, Calif.: RAND, MR-780-A/DARPA, 1996.

Glenn, Russell W., *Marching Under Darkening Skies: The American Military and the Impending Urban Operations Threat*, Santa Monica, Calif.: RAND, MR-1007-A, 1998.

Glenn, Russell W., *". . .We Band of Brothers,"* Santa Monica, Calif.: RAND, DB-270-A, 1999.

Glenn, Russell W., et al., *Denying the Widow-Maker*, Santa Monica, Calif.: RAND, CF-143-A, 1998.

Gordon, Michael, "NATO Air Attacks on Power Plants Pass a Threshold," *The New York Times*, May 4, 1999, p. A1.

Gordon, Michael, and Bernard Trainor, *The Generals' War*, Boston: Little, Brown, 1994.

Gouré, Leon, *The Siege of Leningrad*, Stanford, Calif.: Stanford University Press, 1962.

Grant, Rebecca, "The Epic Little Battle of Khafji," *Air Force Magazine*, February 1, 1998, pp. 28–34.

Granville, Amos R., CO, HMM-261, Oral History Interview, Washington, D.C.: Marine Corps Historical Center, Marine Corps Oral History Program, May 22, 1984.

Greenspan, R., et al., *Robust Navigation Panel Final Report*, Cambridge, Mass.: Draper Laboratory, Report CSDL-R-2833, 1998.

Hacaoglu, Selcan, "U.S. Air Force Using Concrete Bombs Against Iraq," *Associated Press Newswires*, October 7, 1999.

Hackworth, David, "Rangers Ambushed in Somalia," *Soldier of Fortune,* Vol. 19, January 1, 1994, pp. 93–94.

Hallion, Richard, *Storm over Iraq: Air Power and the Gulf War,* Washington, D.C.: Smithsonian Institution Press, 1992.

Hammel, Eric, *Fire in the Streets: The Battle for Hue Tet 1968,* Pacifica, Calif.: Pacifica Press, 1991.

Harden, Elaine, and John M. Broder, "Clinton's War Aims: Win the War, Keep the U.S. Voters Content," *The New York Times,* May 22, 1999, p. A1.

Harden, Elaine, and Steven Lee Myers, "Bombing United Serb Army As It Debilitates Economy; Yugoslav Rift Heals, NATO Admits," *The New York Times,* April 30, 1999, pp. A1, A13.

Harwick, Jon T., CO HMM 365, *Operation Silver Wake,* Oral History Interview, Washington, D.C.: Marine Corps Historical Center, Marine Corps Oral History Program, June 12, 1997.

Hayward, Joel S. A., "Stalingrad: An Examination of Hitler's Decision to Airlift," *Airpower Journal,* Vol. 11, Spring 1997, pp. 21–37.

Headquarters, 1st Battalion, 5th Marines, 1st Marine Division, "Command Chronology for Period 1–31 March 1968," Washington, D.C.: Marine Corps Historical Center.

Headquarters, 50th Troop Carrier Wing, Office of the A-2, "Analysis of Bastogne Resupply by Units of This Command," Maxwell AFB, Ala.: U.S. Air Force Historical Research Agency, January 18, 1945.

Headquarters, Mediterranean Allied Air Forces (MAAF), "Air and Ground Lessons from the Battle of Cassino, March 15–27, 1944," Maxwell AFB, Ala.: U.S. Air Force Historical Research Agency, May 4, 1944.

Henderson, D., "Lidar System Finds Fault with Trees," *Photonics Spectra,* August 1999, pp. 22–24.

Henkel, S., "Tunable Electronic Nose Measures Increased Resistance of Expanding Elements," *Sensors,* March 1999, p. 6.

Henkin, Louis, *How Nations Behave: Law and Foreign Policy*, New York: Praeger, 1968.

Hewish, Mark, et al., "Ultra-Wideband Technology Opens Up New Horizons," *Jane's International Defense Review*, No. 2, 1999, pp. 20–22.

Hibbert, Christopher, *The Battle of Arnhem*, London: Batsford, 1962.

Hicks, J., "Genetics and Drug Discovery Dominate Microarray Research," *R & D Magazine*, February 1999, pp. 28–33.

Horn, S., et al., "Third Generation Sensors," *Proceedings of the IRIS Specialty Group on Passive Sensors*, 1999, Vol. 1, pp. 403–415.

Hosmer, Stephen T., *Constraints on U.S. Strategy in Third World Conflicts*, New York: Crane Russak, 1987.

Hughes, Thomas Alexander, *Over Lord: General Pete Quesada and the Triumph of Tactical Air Power in World War II*, New York: The Free Press, 1995.

Humphries, John G., "Operations Law and the Rules of Engagement," *Airpower Journal*, Vol. 6, No. 3, Fall 1992, pp. 25–41.

Hussain, M., "Ultra-Wideband Impulse Radar—An Overview of the Principles," *IEEE AES Systems Magazine*, September 1998, pp. 9–14.

"Integrated Acoustic Sensors for RFPI," *Proceedings of the 5th Battlefield Acoustics Symposium*, Ft. Meade, Md., September 23–25, 1997, pp. 326–357.

Jenkins, Simon, "NATO's Moral Morass," *The Times*, London, April 28, 1999.

Johnson, P., et al., "Micro Aerial Vehicle Communications Architecture for Urban Operations," *AUVSI '98*, Association for Unmanned Vehicle Systems International, 1998.

Jones, Timothy A., *Attack Helicopter Operations in Urban Terrain*, Ft. Leavenworth, Kansas: U.S. Army, Command and General Staff College, December 20, 1996.

Karnow, Stanley, *Vietnam: A History*, New York: Penguin, 1997.

Keaney, Thomas A., and Eliot A. Cohen, *Gulf War Air Power Survey Summary Report*, Washington D.C.: Department of the Air Force, 1993.

Kennedy, Kevin, *MOUT: An Airman's Perspective*, briefing to a conference on "The Role of Aerospace Power in Joint Urban Operations," sponsored by Air Combat Command, Air Force Special Operations Command, and Headquarters USAF (AF/XPX), Hurlburt Field, Fla., March 23–24, 1999.

Kile, F., et al., "Enhanced Recognition and Sensing LADAR (ERASER) Long Range 2-D Imaging," *Proceedings of the IRIS Specialty Group on Active Systems,* 1998, Vol. 1, pp. 19–31.

Kimm, Peter M., Director, Office of Housing and Urban Programs, Agency for International Development, Testimony before House Select Committee on Hunger, Washington, D.C., November 14, 1991.

Knoth, Artur, "Disabling Technologies: A Critical Assessment," *International Defense Review,* July 1994, pp. 33–39.

Kohn, Richard H., and Joseph P. Harahan, eds., *Air Interdiction in World War II, Korea, and Vietnam*, Washington, D.C.: U.S. Air Force, Office of Air Force History, 1986.

Kokoski, Richard, "Non-Lethal Weapons: A Case Study of New Technology Developments," *SIPRI Yearbook 1994,* Oxford University Press, 1994, pp. 367–386.

Komarov, Alexei, "Chechen Conflict Drives Call for Air Force Modernization," *Aviation Week & Space Technology,* February 14, 2000.

Krulak, Charles C., "The United States Marine Corps in the 21st Century," *RUSI Journal,* August 1996.

Kuehl, Daniel T., "Airpower vs. Electricity: Electric Power As a Target for Strategic Air Operations," *Journal of Strategic Studies*, Vol. 18, No. 1, March 1995, pp. 237–266.

Lambeth, Benjamin S., *Russia's Air Power in Crisis*, Washington, D.C.: Smithsonian Institution Press, 1999.

Larson, Eric V., *Casualties and Consensus: The Historical Role of Casualties in Domestic Support for U.S. Military Operations*, Santa Monica, Calif.: RAND, MR-726-RC, 1996.

Latimer, John C., "Considerations for Operations on Urban Terrain by Light Forces," Ft. Leavenworth, Kansas: U.S. Army Command and General Staff College, Master's thesis, 1985.

Lawson, Mark, "Flattening a Few Broadcasters," *Guardian* (London), April 24, 1999, p. 18.

Leachtenauer, J., "National Imagery Interpretability Rating Scales: Overview and Product Description," *ASPRS/ASCM Annual Convention and Exhibition Technical Papers: Remote Sensing and Photogrammetry*, Vol. 1, 1996, pp. 262–272.

Leachtenauer, J., et al., "General Image Quality Equation: GIQE," *Applied Optics*, Vol. 36, 1997, pp. 8322–8328.

Lederer, Edith M., "Tuzla Off Limits to Off-Duty U.S. Troops," *Detroit News*, February 20, 1997, p. A12.

Lenorovitz, Jeffrey M., "U.N.-Sponsored Airlift to Sarajevo Succeeds in Face of Frequent Attacks," *Aviation Week & Space Technology*, July 27, 1992, pp. 60–61.

Lewotsky, K., "Mars Surveyor Altimeter Flies High in Orbital Test," *Laser Focus World*, November 1997, pp. 43–46.

Lewy, Guenter, *America in Vietnam*, New York: Oxford University Press, 1978.

Lievan, Anatol, *Chechnya: Tombstone of Russian Power*, New Haven and London: Yale University Press, 1998.

Linden, Eugene, "The Exploding Cities of the Developing World," *Foreign Affairs*, January–February 1996, pp. 52–65.

Lorenz, Frederick M., "'Less-Lethal' Force in Operation United Shield," *Marines Corps Gazette*, September 1995.

Mann, Paul, "Strategists Question U.S. Steadfastness," *Aviation Week & Space Technology*, August 31, 1998, p. 32.

Mark, Eduard, *Aerial Interdiction: Air Power and the Land Battle in Three American Wars*, Washington, D.C.: Center for Air Force History, 1994.

Marshall, S. L. A., *Bastogne: The First Eight Days*, Maxwell AFB, Ala.: U.S. Air Force Historical Research Agency, 1988.

Marsili, R., "Lab-on-a-Chip Poised to Revolutionize Sample Prep," *R & D Magazine*, February 1999, pp. 34–38.

Martinez, R. Sandia, Albuquerque, N. M., National Laboratories, Private Communication, February 13, 1998.

Matheson, Michael J., "The United States' Position on the Relation of Customary International Law to the 1977 Protocols Additional to the 1949 Geneva Conventions," *American University Journal of International Law and Policy*, Vol. 2, 1987, pp. 419–431.

McConnell, Malcolm, *Just Cause: The Real Story of America's High-Tech Invasion of Panama*, New York: St. Martin's Press, 1991.

MacDonald, Charles B., *A Time for Trumpets: The Untold Story of the Battle of the Bulge*, New York: William Morrow, 1985.

McKee, Alexander, *Caen: Anvil of Victory*, New York: St. Martin's Press, 1964.

McKee, Alexander, *The Race for the Rhine Bridges, 1940, 1944, 1945*, New York: Stein and Day, 1971.

McLaurin, R. D., and Paul A. Jureidini, *The Battle of Beirut, 1982*, Aberdeen Proving Ground, Md.: U.S. Army Human Engineering Laboratory, January 1986.

McPeak, Merrill A., *Presentation to the Commission on Roles and Missions of the Armed Forces*, Washington, D.C.: Headquarters United States Air Force, September 14, 1994.

Medlin, Larry, CO, HMM-162, Oral History Interview, Washington, D.C.: Marine Corps Historical Center, Marine Corps Oral History Program, November 20, 1983.

Meyerowich, Drew R., Commander, Alpha Company, 2/14 Infantry, 10th Mountain Division, Interview, U.S. Army Center of Military History, April 18, 1994.

Middle East Watch, *Needless Deaths in the Gulf War: Civilian Casualties During the Air Campaign and Violations of the Laws of War*, New York, 1991.

Miller, L. S., "Counter Sniper Technology," *Proceedings of the 5th Battlefield Acoustics Symposium*, Ft. Meade, Md., September 23–25, 1997, pp. 681–692.

"Mini Electronics Smarten Up Small Units," *Jane's International Defense Review*, No. 8, 1998, pp. 34–35.

Mooney, J., et al., "Robust Target Identification in White Gaussian Noise for Ultra-Wideband Radar Systems," *IEEE Transactions on Antennas and Propagation*, Vol. 46, No. 12, December 1998, pp. 1817–1823.

Morin, Richard, "Poll Shows Most Americans Want Negotiated Settlement," *Washington Post*, May 18, 1999, p. A18.

Moroz, S., et al., "Airborne Deployment of and Recent Improvements to the Viper Counter Sniper System," *Proceedings of the IRIS Specialty Group on Passive Sensors*, Vol. 1, 1999, pp. 99–106.

Moser, P., et al., "Complex Eigenfrequencies of Axisymmetric Perfectly Conducting Bodies: Radar Spectroscopy," *Proceedings of the IEEE*, Vol. 71, No. 1, January 1983, pp. 171–172.

Mueller, John, *Policy and Opinion in the Gulf War*, Chicago: University of Chicago Press, 1994.

Muellner, George K., "Technologies for Air Power in the 21st Century," paper presented at a conference on "Air Power and Space—Future Perspectives" sponsored by the Royal Air Force, Westminster, London, England, September 12-13, 1996.

Murphy, John R., "Memories of Somalia," *Marine Corps Gazette*, April 1998, pp. 22–23.

Murray, William, with Wayne W. Thompson, *Air War in the Persian Gulf,* Baltimore, Md.: The Nautical and Aviation Publishing Company of America, 1995.

Myers, Steven Lee, "All in Favor of This Target, Say Yes, Si, Oui, Ja," *The New York Times,* April 25, 1999, Sec. 4, p. 1.

NATO, "Frontal and Army Aviation in the Chechen Conflict," gopher://marvin.nc3a.nato.int/00/secdef/csrc/adv1020%09%09%2B, December 19, 1995 (downloaded November 13, 1998).

Niderost, Eric, "Gallant Defense at Arnhem Bridge," *World War II,* Vol. 11, January 1, 1997, p. 81.

Odom, Thomas P., *Dragon Operations: Hostage Rescues in the Congo, 1964–1965,* Ft. Leavenworth, Kansas: U.S. Army Command and General Staff College, Combat Studies Institute, 1988.

Office of the Under Secretary of Defense for Acquisition and Technology, *Report of the Defense Science Board Task Force on Military Operations in Built-Up Areas,* Washington, D.C., November 1994.

Operation Just Cause Lessons Learned: Volume II: Operations, Ft. Leavenworth, Kansas: Center for Army Lessons Learned, October 1990.

Optech Inc., "Airborne Laser Terrain Mapper," URL: http://home.ica.net/~esk/altm.html, August 1999.

OSD/DARPA UWB Radar Review Panel, *Assessment of Ultra-Wideband (UWB) Technology,* Washington, D.C., Report R-6280, July 13, 1990.

O'Toole, Fintan, "NATO's Actions, Not Just Its Cause, Must Be Moral," *Irish Times,* April 24, 1999, p. 11.

Page, E., "The SECURES Gunshot Detection and Localization System, and Its Demonstration in the City of Dallas," *Proceedings of the 5th Battlefield Acoustics Symposium,* Ft. Meade, Md., September 23–25, 1997, pp. 693–716.

Parker, T. W., "Operation Sharp Edge," *U.S. Naval Institute Proceedings,* Vol. 117, May 1991, pp. 103–106.

Parkerson, John Embry, Jr., "United States Compliance with Humanitarian Law Respecting Civilians During Operation Just Cause," *Military Law Review*, Vol. 133, 1991, pp. 31–140.

Parks, W. Hays, "Air War and the Law of War," *Air Force Law Review*, Vol. 32, 1990.

Pentagon Papers (Gravel Edition), Vol. IV, Boston: Beacon Press, n.d.

Perlez, Jane, "Serbia Shuts 2 More Papers, Saying They Created Panic," *The New York Times*, October 15, 1998.

Peters, Ralph, "The Future of Armored Warfare," *Parameters*, Autumn 1997,

———, "Our Soldiers, Their Cities," *Parameters*, Spring 1996, pp. 43–49.

Plaster, John L., *The Ultimate Sniper: An Advanced Training Manual for Military and Police Snipers*, Boulder, Colo.: Paladin Press, 1993.

Priest, Dana, and William Drozdiak, "NATO Struggles to Make Progress from the Air," *Washington Post*, April 18, 1999, p. A1.

"Pseudolites—A GPS Jamming Countermeasure?" *Flight International*, July 28–August 3, 1999.

"Radar Flashlight Illuminates Humans Behind Walls," *Signal*, June 1998, pp. 89–90.

Reed, Ronald M., "Chariots of Fire: Rules of Engagement in Operation DELIBERATE FORCE," in Robert C. Owen, ed., *Deliberate Force: A Case Study in Effective Air Campaigning: Final Report of the Air University Balkans Air Campaign Study*, Maxwell AFB, Ala.: Air University Press, December 1999.

Reisman, W. Michael, "The Lessons of Qana," *Yale Journal of International Law*, Vol. 22, 1997, pp. 381–399.

Rosenau, William G., "Every Room Is a New Battle: The Lessons of Modern Urban Warfare," *Studies in Conflict and Terrorism*, Vol. 20, 1997, pp. 371–394.

Rosenberg, Barbara Hatch, "'Non-lethal' Weapons May Violate Treaties," *The Bulletin of the Atomic Scientists,* September/October 1994, pp. 44–45.

Sachtleben, Glen R., "Operation SHARP EDGE: The Corps' MEU (SOC) Program in Action," *Marine Corps Gazette,* Vol. 75, November 1991, pp. 77–86.

Sallager, F. M., *Operation STRANGLE (Italy, Spring 1944): A Case Study of Tactical Air Interdiction,* Santa Monica, Calif.: RAND, R-0851-PR, February 1972.

Savage, N., "Lidar Sensor Sees Forest and the Trees," *Laser Focus World,* May 1999, pp. 71–72.

Scales, Robert H., "The Indirect Approach: How U.S. Military Forces Can Avoid the Pitfalls of Future Urban Warfare," *Armed Forces Journal International,* October 1998.

Schiff, Ze'ev, and Ehud Ya'ari, *Israel's Lebanon War,* New York: Simon and Schuster, 1984.

Schindler, Dietrich, and Jiri Toman, eds., *The Laws of Armed Conflicts,* The Hague, Netherlands: Martinus Nijhoff, 1988.

Schneider, Greg, *Nonlethal Weapons: Considerations for Decision Makers,* Urbana-Champaign: University of Illinois, ACDIS Occasional Paper, 1997.

Scholtz, R., "Multiple Access with Time-Hopping Impulse Modulation," *Proceedings of IEEE MILCOM '93,* Boston, Mass., October 11–14, 1993.

Scott, W., "UWB Industry Fate May Hinge on Review," *Aviation Week & Space Technology,* December 14, 1998, pp. 63–64.

———, "UWB Technologies Show Potential for High-Speed, Covert Communications," *Aviation Week & Space Technology,* June 4, 1990, pp. 40–44.

Siegel, Adam B., *Eastern Exit: The Noncombatant Evacuation Operation (NEO) from Mogadishu, Somalia, in January 1991,* Alexandria, Va.: Center for Naval Analyses, 1992.

Spector, Ronald H., *U.S. Marines in Grenada*, Washington, D.C.: Headquarters, U.S. Marine Corps, History and Museum Division, 1987.

Stephens, Alan, *Kosovo, Or the Future of War*, RAAF Fairbairn, Australia: Royal Australian Air Force, Air Power Studies Center, Paper Number 77, August 1999.

Stevenson, Jonathan, *Losing Mogadishu: Testing U.S. Policy in Somalia*, Annapolis, Md.: Naval Institute Press, 1995.

Studt, T., "Development of Microfluidic UHTS Systems Speeding Up," *R & D Magazine*, February 1999, p. 43.

Su, F., "Surveillance Through Walls and Other Opaque Materials," *OE Reports*, No. 140, October 1995, pp. 1–3.

Taw, Jennifer Morrison, *Operation Just Cause: Lessons for Operations Other Than War*, Santa Monica, Calif.: RAND, MR-569-A, 1996.

Taw, Jennifer Morrison, and Bruce Hoffman, *The Urbanization of Insurgency: The Potential Challenge to U.S. Army Operations*, Santa Monica, Calif.: RAND, MR-398-A, 1994.

"Time Modulated–Ultra Wideband Radio Measurement and Spectrum Management Issues," URL: http://www.time-domain.com.

Todd, Oliver, *Cruel April: The Fall of Saigon*, trans. Stephen Becker, New York and London: W. W. Norton, 1990.

Tubbs, James O., *Beyond Gunboat Diplomacy: Forceful Applications of Airpower in Peace Enforcement Operations*, Maxwell AFB, Ala.: Air University Press, 1997.

Tugwell, Maurice, *Arnhem: A Case Study*, London: Thornton Cox, 1975.

U.S. Air Force, Headquarters PACAF, *Airlift to Besieged Areas, April 7–August 31 1972: Project CHECO Southeast Asia Report*, Maxwell AFB, Ala.: U.S. Air Force Historical Research Agency.

U.S. Air Force, Headquarters PACAF, *The Battle for An Loc, 5 April–26 June 1972: Project CHECO*, Hickam AFB, Hawaii, January 31, 1973.

U.S. Air Force, Office of PACAF History, *The Fall and Evacuation of South Vietnam*, Maxwell AFB, Ala.: U.S. Air Force Historical Research Agency, April 30, 1978.

U.S. Department of the Air Force, *International Law—The Conduct of Armed Conflict and Air Operations*, Washington, D.C.: Air Force Pamphlet 110-31, November 1976.

U.S. Department of the Army, *An Infantryman's Guide to Combat in Built-Up Areas*, Washington, D.C.: FM 90-10-1, October 1995.

————, *Operations*, Washington, D.C.: Field Manual 100-5, May 1976.

U.S. Department of Defense, *Conduct of the Persian Gulf War*, Final Report to Congress, Washington, D.C.: GPO, 1992.

U.S. Department of the Navy, *The Commander's Handbook on the Law of Naval Operations*, NWP1-14M/FMFM 1-10/COMDTPUB P5800.7, October 1995.

U.S. Marine Corps, *Aviation Combat Element: Military Operations on Urban Terrain Manual*, MCAS Yuma, Ariz.: MAWTS-1, Edition VII, August 1998.

U.S. Marine Corps, Combat Development Command, *A Concept for Future Military Operations on Urban Terrain*, Quantico, Va., July 25, 1997.

Vallance, Andrew G. B., *The Air Weapon: Doctrines of Air Power Strategy and Operational Art*, New York: St. Martin's Press, 1996.

Van Creveld, Martin, *The Sword and the Olive: A Critical History of the Israeli Defense Force*, New York: Public Affairs, 1998.

Van Creveld, Martin, with Steven L. Canby and Kenneth S. Brower, *Air Power and Maneuver*, Maxwell AFB, Ala.: Air University Press, July 1994.

Venter, A., "Trials Planned for Artificial 'Dog's Nose'," *Jane's International Defense Review*, No. 3, 1999.

Vick, Alan, et al., *Enhancing Air Power's Contribution Against Light Infantry Targets*, Santa Monica, Calif.: RAND, MR-697-AF, 1996.

Vogel, Steve, "Bosnia Airdrop Crews Glad for Dark, Clouds," *Air Force Times*, March 15, 1993, p. 4.

———, "Provide Promise Takes Supplies to Sarajevo," *Air Force Times*, Vol. 52, July 20, 1992, p. 16.

Wagoner, Fred E., *Dragon Rouge: The Rescue of Hostages in the Congo*, Washington, D.C.: National Defense University Research Directorate, 1980.

Warden, John A., III, "The Enemy As a System," *Air Power Journal*, Vol. IX, No. 1, Spring 1995.

Waxman, Matthew C., "Coalitions and Limits on Coercive Diplomacy," *Strategic Review*, Vol. 25, No. 1, Winter 1997.

Weigley, Russell F., *The American Way of War*, Bloomington: Indiana University Press, 1977.

Williams, Mike, "Battle for Khafji," *Soldier of Fortune*, May 1, 1991, pp. 48–52.

Withington, P., "Impulse Radio Overview," URL http://www.time-domain.com.

———, "In-Building Propagation of Ultra-Wideband RF Signals," URL: http://www.time-domain.com.

World Resources 1998–1999: A Guide to the Global Environment, Oxford: Oxford University Press, 1998.

Wright, Brooks, "Urban Close Air Support: The Dilemma," *USAF Weapons Review*, Summer 1998.

Zebker, H., et al., "Topographic Mapping from Interferometric Synthetic Aperture Radar Observations," *Journal of Geophysical Research*, Vol. 91, April 1996, pp. 4993–4999.